The Guide to WorkFace ~~Planning~~ S0-ALM-039

A couple of years ago when Lloyd Rankin and Geoff Ryan started to think about writing this book they envisioned a book that would introduce the application of WorkFace Planning to a world hungry for change. The book would cover What and Why and When and How and Where and Who. It would describe the concepts in broad terms and give concrete examples of the day-to-day activities. Two years down that road, they now understand that the process has many stakeholders, with unique needs that cannot be satisfied with one book.

The outcome is that they have decided to initially create two books, the concept: *"You Can Have it All"* and the concrete: *"Schedule for Sale"*. Ryan has developed this book "Schedule for sale" with input and guidance from Rankin and now Rankin will create "You can have it all" with input and guidance from Ryan. The creation of two books from different perspectives is aimed at developing an effective product that is fit for purpose….. your purpose.

The authors of both books expect that every creation of WorkFace Planning will travel through a conceptual stage followed by concrete application. The books are designed to guide the readers through each of these stages.

As a result of developing two books on the same subject, they expect that you will find the logic and ideas are overlapping and interwoven between the books. This was created by design so that the readers of either book will have a complete understanding of WorkFace Planning, albeit from different perspectives.

You Can Have it All weaves the application of WorkFace Planning with executive management and safety concepts across the breadth of the whole construction industry.

Schedule for Sale details the step-by-step application of WorkFace Planning for project management teams based upon Ryan's experience with Industrial Oil and Gas projects in North America.

Schedule for Sale

WorkFace Planning for Construction Projects

Geoff Ryan P.M.P.

authorHOUSE®

AuthorHouse™
1663 Liberty Drive
Bloomington, IN 47403
www.authorhouse.com
Phone: 1-800-839-8640

First published by AuthorHouse   11/23/2009

ISBN: 978-1-4490-4195-3 (e)
ISBN: 978-1-4490-4196-0 (sc)
ISBN: 978-1-4490-4197-7 (hc)

Library of Congress Control Number: 2009911298

Printed in the United States of America
Bloomington, Indiana

This book is printed on acid-free paper.

Schedule for Sale

The Title:

Can you really buy time on a construction project?

Yes you can, it's called WorkFace Planning and the price is effort and conviction.

The return is the confidence that your project took the time that it had to, and cost the right amount.

Each construction project is governed by the golden triangle of Quality, Cost and Schedule. This is a principle from the Project Management Institute that is present on simple weekend projects and multi-billion dollar projects. A Plumber once summed it up for me by telling me that I could have a project:

Good and Fast, but it won't be Cheap.

Or Good and Cheap, but it won't be Fast.

Or Fast and Cheap, but it won't be Good

In a world where production is money, the industrial and commercial projects that we build cannot afford to cut the corner on Quality, so we have established Quality as the governing influence. That leaves us a choice between Schedule and Cost for the one that we will let slip. The truth is that time is money when you have to hire people to do stuff, so if you want to reduce costs then you have to minimize the schedule. So the Plumbers analogy is not correct, you can have it good, fast and cheap, you just have to keep your focus on good and fast. WorkFace Planning is a process that improves quality and reduces construction costs by minimizing schedules through improved productivity.

The idea that Schedule (productivity) is "for sale" suggests that good productivity is not naturally occurring and that we have a choice between good and bad productivity. We do, and most often, we make the passive choice to leave productivity behind because we don't understand the link between preparation and productivity. This is true at the field level, amongst middle management and in the early design stages of a project. In the wildly complicated world of mega projects our productivity has very little to do with field level desire and is mostly governed by preparation and planning.

WorkFace Planning does not actually "buy" schedule, it is a trade off that exchanges effort and conviction, (preparation), for a risk mitigation strategy that prevents the loss of schedule.

So WorkFace Planning does not allow you to "buy back" the schedule that you lost through poor planning; it prevents you from losing it in the first place.

And it works.

The Concept:

WorkFace Planning is a process that was identified as a best practice amongst constructors by the Construction Owners Association of Alberta (COAA) in 2005. This was the result of a quest by COAA to capture the essence of construction productivity that started with the development of a subcommittee in 2001.

By the end of the year 2000, several high profile mega projects had blown their estimates by billions of dollars and the economic environment was driving the development of many more multi-billion dollar projects. The industry was waking up to the fact that we didn't know how to manage fast track mega projects. The game was changing and we were still trying to apply the old rules.

Lloyd Rankin and I joined the COAA WorkFace Planning committee on the same day in October 2003. Lloyd as a PhD student from the University of Calgary and myself as a representative of KBR and Syncrude. Over the next couple of years we worked together to conduct over 100 interviews with industry professionals and were major contributors to the development of the COAA model for WorkFace Planning.

Our search of the industry revealed some pockets of high productivity and a closer look at these organizations showed us detailed plans built by experienced constructors with a process for removing constraints. This became the foundation for the COAA model for WorkFace Planning, which can be viewed at: www.workfaceplan.com

In 2006, the real world application of WorkFace Planning gave us a completely new level of understanding. The need to capture this experience and expand the influence of this concept across the industry has led me to develop this guidebook.

My company, Insight-wfp.inc, has now established WorkFace Planning on 6 large projects each over $300 million, across both Canada and the US, hired and trained over 50 WorkFace Planners and is consulting for Owners, Engineers, Project Management Teams, Constructors and Software developers. This experience, along with the foundation of the COAA model, is the basis for Schedule for Sale.

To add another level of value to the book I presented the first draft to a wide selection of industry professionals through an online website where readers were invited to read a page at a time and then make comments on that page. I have incorporated their comments and it has enriched both the real world validity of the contents and my own understanding of the issues that we face. Many thanks to all of the contributors.

The Book:

The book will guide you through the Basic Concepts of WorkFace Planning and Resource Requirements to Information Streams and finally to Total Information Management.

The website www.scheduleforsale.com comes as an extension to the book that will allow you to view all of the documents that are illustrated in the chapters. Membership to the site will allow you utilize the interactive flowchart and then download a complete set of sample documents. Members will also have access to a chat room with other readers, where ideas can be floated and tested against the global experience of your peers.   An electronic copy of the book is also available for purchase from the site.

My expectation is that this book combined with the website, your existing knowledge and some other links provided on the back page, will allow you to develop a complete understanding and appreciation of WorkFace Planning. I also believe that this book contains enough of the right tools, concepts and concrete examples to allow you to integrate WorkFace Planning into your existing organization.

With or without further help from us.

Geoff Ryan

Disclaimer:

A mind once stretched by a new idea never regains its original dimensions.

Contents

Introduction:

Overview: WorkFace Planning

The basic concept of WorkFace Planning, or WFP, is that we can reduce our construction schedules by improving the coordination of information, tools and materials at the work face, where the work is performed. This is accomplished by developing detailed, achievable plans that are based upon reality and experience.

The primary product of this effort is a Field Installation Work Package (FIWP).

Historically the Planners in every project organization are a long way from the workface, are over burdened and quite often lack the experience of the people who are actually doing the work. This created large plans that the workforce could not execute in an optimal way. The solution that the industry is migrating towards is to move the detailed planning function closer to the work face, develop planners who understand the work, have enough of them that they can build quality plans, (1 Planner per 50 craft) and then give them direct access to all of the project's information.

= WorkFace Planners

These two elements are the basic principles of WFP, i.e. Planners and Plans.

Let's have a closer look at the details of each one.

Chapter 1

FIWPs: (Field Installation Work Packages)

A sample FIWP is available under the sample documents folder of the website.

http://www.scheduleforsale.com

A FIWP is a detailed plan that contains 500 to 1000 hours of work that will be executed during one rotation, by one crew, from a single trade discipline. Based on 10 workers and a Foreman working 10-hour shifts for a rotation of either 5 days (500hours) or 10 days (1000 hours). The expectation is that the crew will start work on their 500-hour plan when their shift starts on Monday and the plan will be complete by the end of the shift on Friday. For the purpose of this book, we will use the 1000-hour packages. The "10 days on / 4 days off" schedule is emerging as the most common shift for our application of industrial construction.

The logic behind this scale is that most of the activities on any work site are based upon the rotation. We start activities with the intent of getting them finished by the end of the rotation or we envision a certain stage of progress before we go home for the weekend.

The Contents of each FIWP:

Each FIWP must have:-

A cover page: that shows a 3D picture of the scope, a one line definition of the scope, the FIWP number, and the Planned Value in work hours.

A table of Contents: This is a basic function of all good technical documents, provided so that the reader does not have to read the whole document to get the answer that they are looking for.

Constraint page: This page will list the constraints in order, showing the status of each one with a final sign off column for QC, Safety and the Superintendent. A note of warning: There is a temptation to get all of the stakeholders to sign off on a FIWP before the work is released. This will effectively choke the process and add weeks to the development cycle. The only signatures required are for Quality Control, Safety and the Superintendent.

A detailed scope of work: the level of detail here will change based upon the experience of your workforce. In Alberta we are privileged to work with a mature workforce that has at least 7 mega projects of experience over the last 10 years and tradesmen that have all graduated from government run apprenticeship programs. We don't need to tell this workforce the details of a simple rigging lift. At the other extreme if we were building a petrochemical plant in a third world country that has had very little economic activity, we should probably include a diagram that shows a clevis, sling and softeners for a steel rigging plan. Ultimately, the level of detail in a FIWP develops based on the feedback that a WorkFace Planner gets from the Foremen.

Safety Planning: The true value from this page comes when the Foreman utilizes it to create and maintain a culture of safety around the scope of work. There are standard safety documents that should be added to every scope of work, however the Safety representative that reviews this FIWP should also use it as a way to focus the Foreman's attention on specific dangers that are present in the work identified. This could be in the form of a toolbox talk that references the type of work being executed or a heads up statement about the latest injury statistics. Ultimately, the Foreman is responsible for the safe execution of work and this section should be used as a means to help them prepare for that. As a two-way communication tool, the FIWP is also alerting the Safety team to the imminent path of construction.

Quality Planning: The Quality Control team should use this interaction with the FIWP as a means to communicate directly with the Foreman on specific activities. The best way to do this is to extract the portion of the Inspection and Test Plan (ITP) that is relevant to the scope and highlights the places where the Foreman will need to get signatures, notify inspectors or implement holds. By adding the documents required for turnover, the QC team can effectively utilize the FIWPs to facilitate turnover from the first day of construction. As the FIWP is returned from the field, the QC team extracts the completed documents and replaces them with copies. The original documents are filed under their system number in the QC office. If there is a problem with the documents, it allows the QC inspectors to address the problem immediately with the Foreman who did the work. By reviewing all of the FIWPs the QC teams are also being kept in the loop for the execution of construction activities.

This section requires a sign off from the Quality Control Representative, to ensure that every package is reviewed.

Trade Coordination: In the world of fast track construction one of the biggest barriers to productive activity is the problem of trade congestion. It is the responsibility of the WorkFace Planner to be aware of the activities of other trades and to mitigate the risks of delays caused by conflicts of access. This page should be used by the WorkFace Planner to communicate the risks and the mitigation strategies with the Foreman. This could be as simple as the names of the other trade foremen in the area and their radio ID. It could be as complex as a schedule for access to an area to allow another crew some time for an overhead lift.

We can facilitate this interaction between the trades by ensuring that all of the WorkFace Planners are in the same location and have regular coordination meetings. Ultimately, the General Superintendent is responsible to resolve trade access conflicts that cannot be negotiated by the Lead WorkFace Planner.

Material Confirmation: This is a critical component of the FIWP and historically the single element that continually prevents productive activity. The WorkFace Planner must be absolutely confident that material is available and that it will be delivered to the site in the days prior to the first day of execution for this FIWP. This section of the FIWP must show:

- A Bill of Materials (BOM) for the FIWP that shows a complete list of every component required to install this FIWP.
- A copy of the confirmation from the WorkFace Planning Material Coordinator that the material is available and that bulk materials have been hard allocated against this FIWP.
- A copy of the Request for Material with a Required On Site (ROS) date.
- Radio or cell phone contact information for the driver that will deliver the material.

All of this information will be supplied to the WorkFace Planner by the WorkFace Planning Material Coordinator.

Scaffold Confirmation: This should be a one page document that shows the request for scaffolds, and the confirmation that they have been erected, complete with contact information for the Scaffold Foreman. Later in this book, we will explore the detail of how a scaffold management database can be utilized to manage all of this data and produce reports.

Construction Equipment Confirmation: Just like the scaffold request, this one page document (from an equipment management database) must show the list of equipment requested and have confirmation that the equipment will be ready for pick up or delivered to site prior to the commencement of work.

Timesheets and Cost codes: This section should contain all of the timesheets that the Foreman will require with predefined cost codes on each timesheet. One of the fringe benefits that we get from having a defined scope of work is that we can also identify exactly which cost codes the work should be charged to. This removes the guesswork that the Foremen typically struggle with and will give us a higher level of confidence in the costs charged against work. The extension of this logic takes us to the naming convention that we use for the FIWP. It should identify the work using the same naming convention as the Work Breakdown Structure (WBS) and then link the cost code structure to the FIWP, (more on this later). The net result is that the time sheets have the FIWP number as part of their cost coding. This gives us a Productivity Factor (PF) for each FIWP (Crew by rotation) with very little effort and a great deal of accuracy.

Drawings and Model shots: The drawings on any construction project are the ruling document, so the accuracy of this section is critical. Typically the model snap shots that should accompany each drawing are for reference only and should be labeled so. The WorkFace Planner has the sole responsibility to ensure that the FIWP has the latest revisions immediately prior to release. The QC team should also utilize this section to collect weld data (by weld stamping) or installation data for Electrical, Instrumentation and Steel.

Value to the Customer

We get good quality FIWPs when we think of the plan as a product that is developed by a supplier to address the needs of a customer. The planner is the supplier and the work crew is the customer. To satisfy the customer's expectations the scope needs to be achievable and supported by a current reality.

Achievable Scope:

The rule that we have used successfully when developing WorkFace Planners was to ask them to envision how much work that they could achieve with one crew, in one rotation, given the ideal set of circumstances. Imagine that nothing will go wrong and develop your scope based on that. Then add another 10% to make sure that you don't run out of work.

This is a very important basis for planning.

- If we plan for things to go <u>wrong</u> they will,
- if we plan for things to go <u>right</u> they will.

So make a conscious decision on which outcome that you want and don't worry too much about reality. (We will create our own reality)

The common model for this type of thought is:

Where are we Now?

Where do we want to be?

How do we get there?

Note: This model <u>does not</u> ask: How far can I get from here? OK that is what I want. This is the same basis that we use for the "zero safety incidents" mentality. It is not about what we <u>expect</u> to happen it is about what we <u>want</u> to happen. We want to have no safety incidents and we want to build plans that get stuff done.

Current Reality:

The WorkFace Planners create a current reality as they remove constraints. Immediately prior to the release of a FIWP, we would expect that the current reality would be that the material is available, the scaffold has been erected, the work is aligned with the schedule etc.

The presence of these two elements, achievable scope and current reality, is expected to lead to a situation where the General Foreman can hand a FIWP to their Foremen, look them in the eye and say "I expect that you will get this done in one rotation".

These two elements: <u>Achievable Scope</u> and <u>Current Reality</u> (the Removal of Constraints) are therefore critical to the successful execution of FIWPs.

WorkFace Planning Software: The market has several very good products that will load a 3D model with installation unit rates and allow a planner or a group of field experts to click on a component and have the model tell us how long it will take to install (amongst other things). So for WorkFace Planners who are trying to visualize work and estimate how much they can get done, this is like the invention of the wheel. My advice to you is to buy one of these products and use it lots, (much more to come).

Constructing your Field Installation Work Packages:

There are many different ways for WorkFace Planners to build FIWPs. The two ends of this scale might look something like this:

1. Lock yourself in a dark room with some rainforest music and invite your Maker to help you develop the perfect FIWPs.

Or at the other end of the scale;

2. Fill a meeting room with Superintendents, General Foremen, Foremen, Schedulers, and WorkFace Planners. Then project the 3D model onto one wall and pull out a single CWP (Single discipline and <40,000 hours). Now ask the room "which piece goes in first". Then use the WorkFace Planning software to build virtual FIWPs based upon the experience in the room and supported by the installation unit rates that are built into the model.

Your own application of WorkFace Planning will probably be somewhere between these points.

An important rule of the universe to inject here is that: <u>Your</u> results are directly related to <u>your</u> effort.

If you take the time to pull very valuable people out of the field, or bring them in on overtime, you will reap the benefits from a common vision and good quality plans, I've seen it.

Chapter 2

Removal of Constraints:

(This is the creation of reality that I referred to earlier)

As each FIWP is created, it is assumed to have this list of constraints:

1. Construction Work Package – Must be Issued For Construction
2. Scheduled – The work must be aligned with the Path of Construction.
3. Engineering Data – Engineered drawings must be issued and available.
4. Prerequisite work – The work that has to happen before this FIWP can be executed must be complete.
5. Materials – Every component must be identified and confirmed onsite.
6. Scaffold – Must be identified, ordered and built fit for purpose.
7. Construction Equipment – Must be identified and confirmed fit for purpose.
8. Tools – There must be clear access to a reliable supply of the right tools.
9. Resources – Qualified trades people must be available with all of the appropriate site training requirements.
10. Quality documentation – The reference documents that will govern the scope (ITPs) must be identified and available.
11. Safety Planning – There must be a program that will support the Foreman's safe application of the work.
12. Access to the work face – The Permits required and congestion from other activities/trades must have been identified.

The work cannot be considered achievable until all of these constraints have been removed. Any exception to this is a deviation from WorkFace Planning that will cost you schedule (and money, reputation, sleep etc).

Timeline for the Removal of FIWP Constraints

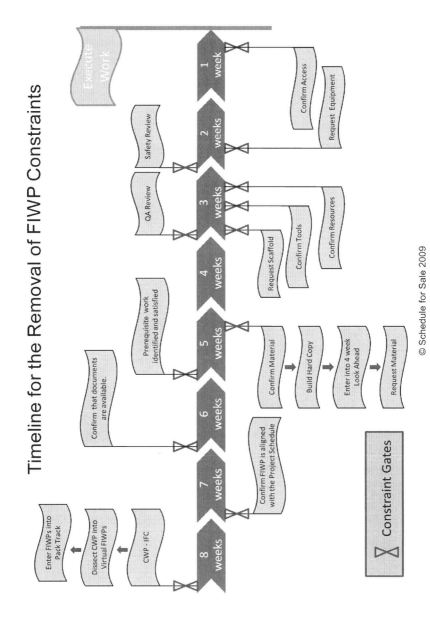

© Schedule for Sale 2009

From birth, the FIWP travels through a series of developmental stages on the path to full maturity and successful execution. As each constraint is satisfied, the FIWP progresses to the next one and so on until we have created a current reality for that package, which allows the scope to be executed.

In the following example you can see 18 FIWPs maturing from left to right and from red to green towards the final column "Progress Recorded" This page represents one Construction Work Package (CWP) for Pipe and shows that we have three FIWPs in the danger zone (red) held up by material and prerequisite work (steel). We have eight FIWPs in the preparation phase, (yellow) and three FIWPs in the chute (green) ready for execution.

It is a very important project management function that this information be tracked and displayed. This single document is the pulse of the project.

A coloured sample of the Pack Track spreadsheet can be downloaded from the website.

PACK TRACK - Constraints Checklist

Pipe : CWP L-400-39

FIWP / STATUS	Level 5 Schedule ID	Level 3 Early Start	Level 3 Late Finish	Progress Recorded (-1)	Work Complete (-1)	RFI Submitted (0)	Package Released (1)	Checked for Latest Revs (1)	Material Request (1)	Scaffold Confirmed (1)	Request Crane & Manlift (2)	Safety Review (2)	Scaffold Requested (3)	QC Review (3)	Material Notification (4)	Super Review (4)	GF Review (4)	Hard Copy Package (4)	Material Received (4)	Equipment in Place (6)	Steel Erected (6)
L-400-39-001-12345	410465A143	Aug23	Sep30				X	X	X	X	X	X	X	X	X	X	X	X	X	X	X
L-400-39-002-12345	410465A142	Aug23	Sep30										X	X	X	X	X	X	X	X	X
L-400-39-003-12345	410465A141	Aug23	Sep30														X	X	X	X	X
L-400-39-004-12345	410465A140	Aug23	Sep30														X	X	X	X	X
L-400-39-005-12345	410465A139	Aug23	Sep30					X	X	X	X	X	X	X	X	X	X	X	X	X	X
L-400-39-006-12345	410465A138	Aug23	Sep30							X	X	X	X	X	X	X	X	X	X	X	X
L-400-39-007-12345	410465A137	Aug23	Sep30														X	X	X	X	X
L-400-39-008-12345	410465A136	Aug23	Sep30									X	X	X	X	X	X	X	X	X	X
L-400-39-009-12345	410465A135	Aug23	Sep30									X	X	X	X	X	X	X	X	X	X
L-400-39-010-12345	410465A134	Aug23	Sep30																	X	X
L-400-39-011-12345	410465A133	Aug23	Sep30										X	X	X	X	X	X	X	X	X
L-400-39-012-12345	410465A132	Aug23	Sep30															X		X	X
L-400-39-013-12345	410465A131	Aug23	Sep30										X	X	X	X	X	X	X	X	X
L-400-39-014-12345	410465A130	Aug23	Sep30															X	X	X	X
L-400-39-015-12345	410465A129	Aug23	Sep30															X	X	X	X
L-400-39-016-12345	410465A128	Aug23	Sep30	X	X													X	X	X	X
L-400-39-017-12345	410465A127	Aug23	Sep30	X	X	X	X	X	X	X	X	X	X	X	X	X	X	X	X	X	X
L-400-39-018-12345	410465A126	Aug23	Sep30	X	X	X	X	X	X	X	X	X	X	X	X	X	X	X	X	X	X

Constraints in Detail:

In the following passages we will take a closer look at what it takes to satisfy each of the constraints. There is another level of detail behind each constraint that we will explore further, later in the book. For now this is a guide that the WorkFace Planners can use to understand how to shape their day to day activities for the removal of constraints.

These activities are based upon the assumption that you have a 3D model and that you have applied a version of WorkFace Planning software. If you do not have a 3D model or have not applied WorkFace Planning software your WorkFace Planners can still function, you will just need more of them.

1. **Construction Work Package (CWP)** – *Issued For Construction (IFC).*

There is a temptation on every project to develop a level-5 detailed schedule as early as possible so that we can have an accurate estimate for exactly how long it will take and how much it will cost. (Please take a moment right now and look up the word "estimate" in the dictionary). The problem is that these two elements: Cost and Schedule, are directly related to quantities. The more you build the longer it takes and the more it costs. So to have any sort of an accurate projection of our Schedule and Cost we must first know how much we have to build. We only achieve this status when engineering releases Construction Work Packages – "Issued for Construction". Any projections developed before this point are truly WAG (Wild Ass Guess) estimates. Please treat them as such.

For the purpose of WorkFace Planning we need accurate information, so the development of FIWPs must not start until a CWP has been released IFC. We will use this rule later in the book to show what a CWP should look like and how the alignment of the Work Breakdown Structure (WBS) will feed into the execution of WorkFace Planning.

For a CWP to be considered IFC, >95% of the drawings must be issued to Document Control. This is one of the cornerstone requirements for the 80/100 rule. 80% of materials received and 100% of the engineering documents must be issued before the work starts.

At the point of IFC the engineering team must also issue the updated 3D model that shows the CWP being released. The drawings delivered to document control must be scanned images (PDF) and the electronic files (PCF, IDF). The WorkFace Planning software will draw these electronic files and build a 3D model from the electronic IFC drawings.

This recreated model can then be compared to the 3D model issued by engineering. This is a very simple way to validate the engineering information (and should be used by the engineering team to check their deliverables prior to release). The WorkFace Planning software works by allowing the user to extract the CWP from the model and colour the portion of the model that has IFC drawings, which will show immediately if any of the scope has not been released.

2. Scheduled – Work is aligned with the Path of Construction.

The Path of Construction:

The level-3 schedule is the ruling document on any construction project. In order to facilitate the development of FIWPs the schedule must be set up to bring together, scope, time and cost. We do this by designing the Work Breakdown Structure (WBS) so that it allows us to dissect the project into:

Construction Work Areas (CWA): A geographical cube of work defined by Construction with less than 100,000 hours. This includes all disciplines, with the exception of cable, tray and undergrounds. Cable, tray and undergrounds are also dissected into CWAs but are across the whole project, not geographically bound. The guiding principle here is to dissect the work so that each discipline could be assigned to a single contractor. This is not the intent but will lead to boundaries that are formed by logical associations of construction work.

Each CWA then becomes one activity on the level-2 schedule

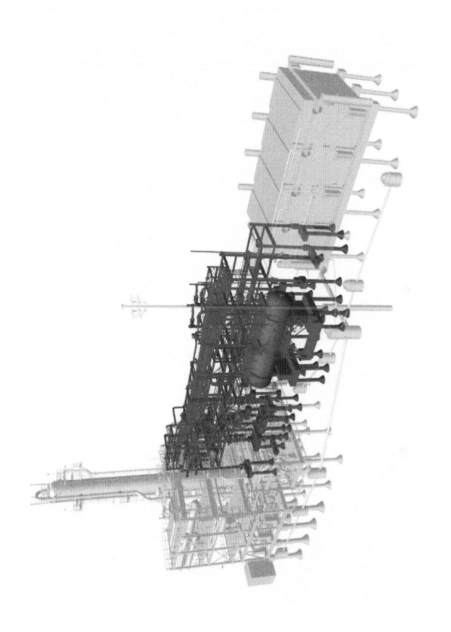

Construction Work Area (CWA)

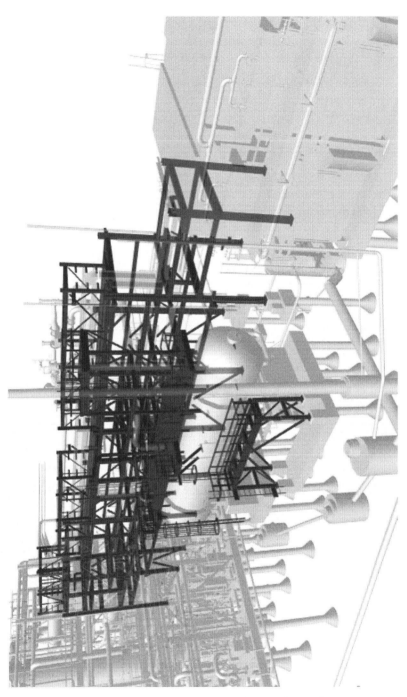

Construction Work Package (CWP)

Construction Work Packages (CWP): Single discipline of a CWA with the same boundaries and less than 40,000 hours. Equal to a single Engineering Work Package (EWP).

Each CWP then becomes one activity on the <u>level-3</u> schedule.

The dissection of work in this format allows the Construction Team to develop the Path of Construction prior to the start of detailed engineering. This activity will produce a schedule of CWPs with lag, dependencies and prerequisites (without duration). The engineering team can then apply resources and time to the activities, which will give them a CWP release plan and allow them to develop a schedule for the development of EWPs.

This gives us a <u>level-3</u> engineering schedule based upon the path of construction.

Take this schedule and write it in stone. The only way to change the Path of Construction is to get consensus from all of the project's stakeholders. (A change that is convenient for one Team is probably detrimental to another).

COAA have a very well documented sample of a process that can be used to generate the Path of Construction on their WorkFace Planning website. <u>www.workfaceplan.com</u>

The CWP release plan is then taken by the Procurement Team, who populate each CWP with the time estimates from their vendors and fabricators. This will give us the start dates for the Construction Schedule by CWP. Based upon the logic that each CWP must be 100% engineered with 80% of the material received prior to the start of construction. (This logic is on its way to becoming common).

The Construction Team then applies resources and time to the CWPs on the <u>level-3</u> Procurement Schedule and this gives us a <u>level-3</u> Construction Schedule and the starting place for WorkFace Planners.

The critical component of this development process is that we must all speak the same language: CWPs. A CWP for engineering is the same as a CWP for procurement and is the same for construction. When we deviate from this, the alignment between scope, time and cost is lost. Work hard to make sure that this does not happen and you will reap great rewards. I've seen it.

To facilitate the correct sequence of fabrication CWPs must not be released by engineering to procurement until they are 100% engineering complete. This will draw the focus of the fabricators in to one CWP at a time. (Which is how we want to build it).

As the CWP is released to the Fabricators, the WorkFace Planners should also receive notification that the CWP is engineered 100% and this will be the trigger for them to dissect the CWP into Virtual FIWPs in the 3D model.

The 3D model administrator will load the IFC CWP into the WorkFace Planning software and then facilitate the FIWP development session: (as described in Chapter 1)

> *Fill a meeting room with Superintendents, General foremen, Foremen, Schedulers, and WorkFace Planners. Then project the 3D model onto one wall and pull out a single CWP (Single discipline and <40,000 hours). Now ask the room "which piece goes in first". Then use the WorkFace Planning software to build virtual FIWPs based upon the experience in the room and supported by the installation unit rates that are built into the model.*

The net result of this session is that we have taken a level-3 activity and broken it into level-5 activities, based upon the way that the field supervisors would like to build it. (This gives us construction smarts built right into the schedule). The Scheduler then takes these components (FIWPs) and enters them into the schedule as level-5 activities dissected from the level-3 activity (The CWP). The actual path of FIWP sequencing within one CWP remains dynamic until the constraints have been satisfied. The four week look-ahead is then developed from the "green" FIWPs.

This process is the application of the **rolling schedule** concept: A level-3 schedule is developed for the entire project, but the level-5 schedule is only developed as we approach the execution of the work. Typically this is 8 - 12 weeks prior to the commencement of the work. This creates a very real schedule that we can expect to be executed as planned. You probably noticed that we skipped over the level-4 schedule, well done. The truth is that we don't need it. If you are using WorkFace Planning software the process is fully functional when we give the WorkFace Planners a level-3 activity and then ask them to dissect it into level-5 activities (more on this later).

When an individual FIWP is developed and entered into Pack Track, the first constraint that it must satisfy is that it must be work that is in line with the path of construction, it must have been derived from the level three construction schedule and it must be in the right sequence. If all three of these conditions have been satisfied then the FIWP can be considered to be "Scheduled".

3. Engineering Data – Drawings with the latest revs.

One of the options that we have with WorkFace Planning software is that after we have built a virtual package in the 3D model we can then get the model to produce a list of drawings sorted by FIWP. This radically simplifies the process of determining which drawings we assigned to a package.

Note: One of the keys to productive activity is to build by whole drawings, rather than by component. We don't plan to put in a single spool of pipe or a single piece of steel, we plan the whole drawing. This does not sound like much, but in the building process if there are components missing and you decide to leave them out and build around them, it adds a whole new level of complication. If you plan by component instead of by drawings you will find that workarounds become the norm and then you spend all of your effort tracking the missing pieces. (Fighting fires instead of planning work).

The WorkFace Planners will pull a list of drawings from the 3D model, and then it is a simple task to visit Document Control and pull the latest drawings. (Yea-right). Ok it's not always simple and you probably have an old revision. I have to admit at this point that we have not yet tamed this beast, but I have seen others do it.

I know of a project outside Canada where the Document Control Database was set up to receive and file documents directly from the engineering team. The filing structure was based upon the naming structure that was established in the 3D model and the database was online. The Planners (in another country, far from the engineering team), hyperlinked the components in the 3D model with the file structure in the database so that you could right click on an object in the model and the latest drawing would appear. This drawing could be 1 day old and is now being printed 100 meters from the workface, 12 time zones away.

The Techy Guys reading this right now are saying "so what, we've been doing this for 10 years". For the seasoned Constructors it is probably more like: "Seeing is almost believing". It's true, it can happen and it will be the normal way of doing business sometime in the future (you decide when). Technology is waiting impatiently for process to catch up.

Right now, you will have to deal with whatever system you have, so it is the WorkFace Planner's sole responsibility to ensure that the drawings that are issued to the field represent the latest revision.

Each drawing must be accompanied by a 3D snapshot that is taken from the same angle (60/30) as the isometric drawings.

Typically, the drawing is the ruling document so it is a good practice to auto stamp the 3D model shots with "Issued for reference only".

Every drawing must be date-stamped on the date that it was drawn from document control and the date that it was last checked for revisions. Typically, the WorkFace Planners do not print the drawings until the package makes the transition from Virtual to Hard copy, after the material constraint has been satisfied.

Each engineering drawing must be supported by the cut sheets that were produced by the fabricator. For steel, this shows the details of connections and for pipe, this shows how the fabricator interpreted the engineering and any changes necessary for shipping.

Engineering data may also include information that is not from a drawing, these could be installation specifications for vendor components, government regulations or site-specific engineering procedures. The WorkFace Planner is responsible to understand the requirements and to communicate the expectations in the FIWP.

In summary: To satisfy this constraint the WorkFace Planner must be confident that the drawings have been Issued For Construction (IFC) and are the latest revision.

4. Prerequisite work – Work that has to happen before a task can be executed.

This constraint is the same as the dependent logic that we establish between scheduled activities. For example, if we are installing a vessel we will need to complete the foundation first. If we are pulling cable, we should have the tray in place before we start the work.

When we are designing this logic, there are many "grey areas" when we think about the sorts of things that absolutely must happen before a certain activity takes place. The deciding factor is the core of WorkFace Planning: Productive Activity.

The price that we pay for being busy but not productive is called "rework". This is work that is being pulled apart and re-routed or pulled out and stored so that another trade or activity can take place.

All good trades people know that the fastest way to do anything is to do it once.

When we sit down to build the Path of Construction during the conceptual stages of a project, we choose the shortest distance between the start line and the finish line. Predetermining the most productive way to build the project. That logic holds until we are in the heat of battle and we run out of cable tray

20 meters before the end of a run. Now we have to decide whether we wait for two more days for the rest of the cable tray or do we go ahead and start pulling cable. The answer is to think about the finish line not the starting line. We can start the race now and run through an obstacle course trying to find the finish line or we can wait for the obstacles to be removed and run in a straight line to the finish line. This is the essence of the difference between Activity and Productive Activity.

On any construction project the builders are not employed to keep busy, they are employed to complete tasks. If this is true then here is the culture shift that we will have to make:

Activity does not always = progress.

To all of the planners reading this the correct answer is to not run out of tray in the first place, but back in the reality of the field, the constructors are probably going to design some type of workaround of scaffold or temporary supports, that will simulate cable tray. The net result is that we have allowed the Electricians to start pulling cable but we have also moved the finish line further away. The correct action in this instance is to stop work until the tray has been completed. (it is not the popular choice but it is the right one).

To satisfy the constraint of prerequisite work the WorkFace Planners need to start with the expectation that all preceding work must be complete before a FIWP is released. This will generate questions like: when do the Ironworkers have to stop erecting steel so that the Boilermakers can set a vessel. Or when has all of the other work been completed so that it is safe to run cable tray to instruments. These are very complex questions that initiate many thousands of conversations. The answers are in those conversations. This is one of the key reasons for housing all of the WorkFace Planners in the same area.

These types of conversations and the logic of Productive Activity are elements of the onsite training that the WorkFace Planners need to help them develop good quality planning skills and executable FIWPs.

For the purpose of developing the list of prerequisite tasks we ask the WorkFace Planner to think through the stage-by-stage construction of the work and identify the dependencies that they will need. This could include other FIWPs from the same discipline or other work from suppliers upstream.

This list of dependencies will appear as a one-page section of the FIWP.

5. Materials – Every component identified and confirmed onsite.

If I was asked to identify the single biggest contributor to poor performance it would have to be the supply of material (followed closely by documents). We will get into the details of material management in a future chapter, for now we will take a leap of faith and assume that the onsite material management group can tell us what they have and what they don't have. (This is a stretch from the current reality).

The deliverable from dividing a CWP into FIWPs is a list of FIWPs and the drawings that belong to each one. The WorkFace Planner can then summarize the Bill of Material from all of the drawings and end up with a complete list of the material required to build a FIWP.

To satisfy this constraint the WorkFace Planner must be able to take that list of material and confirm that it has been received at the lay down yard by the material management group and that it can be delivered to the workface.

Here is another great argument for automated WorkFace Planning software, imagine how long it takes and how many mistake s are made during the manual extraction of a material list from a stack of 20 drawings, (1 FIWP). Automation can give us that list with 100% accuracy and a picture of what it looks like at the press of a few buttons.

It is the WorkFace Planner's responsibility to generate this list for every FIWP. One of the fundamentals for WorkFace Planning is that we must know that we can execute the work before we release the FIWP. For the material constraint to be satisfied this means that we must be confident that we can have the material <u>before</u> we request it. The most effective way to do this is to appoint a WorkFace Planning Material Coordinator.

The Coordinator will then receive the list of FIWPs and the materials that they require from the WorkFace Planners and build a report that shows the materials received and the shorts for each FIWP. This report will change every day based upon the materials received in the lay down yard and the dynamic allocation of bulk material based on priorities.

Material Management to Support WorkFace Planning

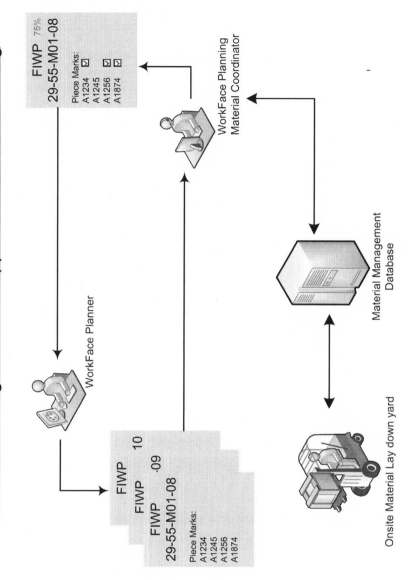

FIWP 75%
29-55-M01-08

Piece Marks:
A1234 ☑
A1245 ☑
A1256 ☑
A1874 ☑

WorkFace Planner

WorkFace Planning
Material Coordinator

Material Management
Database

Onsite Material Lay down yard

FIWP 10
FIWP .09
FIWP
29-55-M01-08

Piece Marks:
A1234
A1245
A1256
A1874

When a FIWP reaches the point where the material constraint has been satisfied it turns yellow in Pack Track and makes the transition from Virtual to Hard copy. This also means that the package is potentially 4 weeks from execution and can be entered into the 4 week look ahead schedule, at the discretion of the Superintendent/General Foreman. The WorkFace Planner will generate the material request when the FIWP is entered into the 4-week look ahead.

The material request will include a Required On Site (ROS) date which should be at least two days prior to the scheduled commencement of the work. This gives the Foreman the chance to organize and check the material and the opportunity to get started early if the previous package is completed ahead of schedule.

The Material Management Team are responsible for the management of information and the handling of all material until it is delivered to the workface. This means that the work crews do not collect their own materials from the laydown area. The logic here is that the people that work in the laydown yard will become very good at what they do (handle material) and the work crews will become very good at what they do (Install material), so we should let them do it.

6. Scaffold – Identified, ordered and built fit for purpose

For the purpose of removing constraints we will assume that the Scaffold management system has the ability to receive and process requests for scaffolds. In the section on scaffold management, we will look closer at exactly how that system works.

Soon after the FIWP makes the transition from "virtual" to "hard copy" the WorkFace Planners will determine the scaffold requirements. The scaffold request will consist of a snap shot from the 3D model with a description of the work to be executed, a required by date and length of service. (An advancement on this in the future will be to request power, lights and/or a fire extinguisher).

The request will be filled out and submitted electronically from an icon on the WorkFace Planners desktop. The WorkFace Planning Scaffold Coordinator will receive the request, electronically, and use it to build FIWPs for the scaffolders. When the work has been completed, the WorkFace Planning Scaffold coordinator will send a confirmation notice to the WorkFace Planner who generated the request. The constraint will be considered satisfied when the WorkFace Planner issues the request, not when the scaffold has been constructed. This means that the scaffold management team then assumes the responsibility to construct every request that has been submitted at least one week prior to the required by date.

This process is on the drawing board to become an automated feature incorporated into several of the WorkFace Planning software tools. The software will allow the user to create scaffolds right in the 3D model and allow the Scaffold Coordinator to zoom in on the point in the model that the scaffold is being constructed to access.

WorkFace Planning Scaffold Request

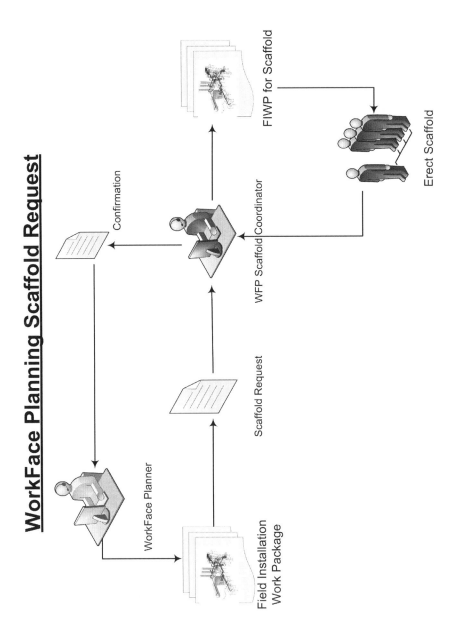

WorkFace Planner

Field Installation
Work Package

Scaffold Request

Confirmation

WFP Scaffold Coordinator

FIWP for Scaffold

Erect Scaffold

Construction Equipment – Identified and confirmed fit for purpose.

The structure of this process is very similar to the scaffold request process; however the way that we currently manage Construction Equipment is a lot further away from this vision of management.

On a typical construction project, we assign cranes, manlifts and welders to a specific Superintendent and they utilize them as best they can. The quantities of these resources is governed by an equipment budget so that if the budget says that we have five cranes and five Superintendents then each Superintendent gets one crane. As their needs fluctuate, they have to beg, borrow or steal resources from the others.

This identifies one of the key constraints that we have to being productive. In an ideal world, we would determine how much work we have to get done and in what time frame, then we would load resources against that model. When we choke off the number of resources, we have effectively changed the schedule to satisfy the equipment budget.

For this reason, the process of equipment allocation must be managed like the scaffold model. The WorkFace Planners will create FIWPs to satisfy the schedule and it is the responsibility of the WFP Equipment Coordinator to satisfy the resource requirements. This is a big change from where we are now.

The reality of the world is that we <u>are</u> constrained by resources so when the WorkFace Planners are building FIWPs we ask them to book cranes by the ¼ day or by a % of use. This will allow the WFP Equipment Coordinator to propose shared equipment, between the trades.

For the purpose of removing the constraint on FIWPs the WorkFace Planner will send an electronic request to the WFP Equipment Coordinator and insert a copy of the request into the FIWP. Then the responsibility lies with the WFP Equipment Coordinator to satisfy the request.

When the Coordinator receives the request, the first thing that we ask them to do is to confirm that the equipment requested is suitable for the task. The equipment request form will ask for the details of weight and distance for cranes and height and ground access for manlifts, along with the Planner's opinion for what size of equipment will work. The WFP Equipment Coordinator will then run their own calculations and determine if the equipment suits the task.

WorkFace Planning Equipment Request

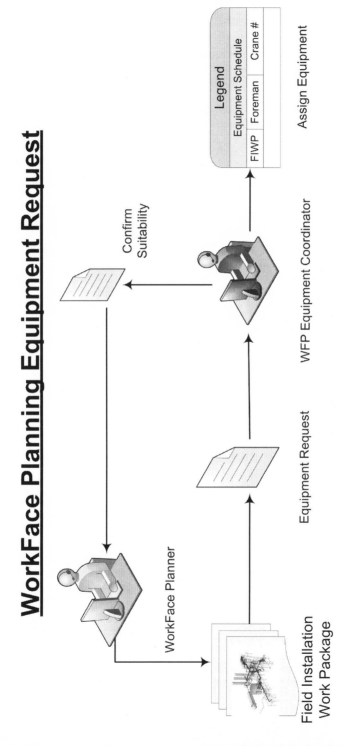

Legend

Equipment Schedule

FIWP	Foreman	Crane #

Assign Equipment

Confirm Suitability

WFP Equipment Coordinator

WorkFace Planner

Equipment Request

Field Installation Work Package

8. **Tools** – Clear access to a reliable supply of the right tools

For many projects this is a normal condition and not something that we have to pay particular attention to. However, we still see the occasional project that has a limited amount of poor quality tools. One recent project in Alberta was an example of this. The contract for supplying tools was based upon a projected workforce and there was no provision for the extra 5000 trades that were hired. So the tool supplier took it upon themselves to choke off the supply of tools. The contract administrator thought the he was doing a good job by sticking to the original terms of the contract but the net result was that a portion of the trades were forced to smuggle their own tools onto site so that they could do the work. The total burdened cost of a trades person at the time was about $100/hr. To pay those people to stand around and wait for a $20 hammer is a very poor business practice. .

The premier tool suppliers in the industry all use Tool Hound or a similar product that scans badges and tools, and then produces reports that show who has what. This report can then be used by the Field Supervisors to monitor for abuse. The sensible tool supply contracts that we see all pay the vendor for tools out, so it is in the vendors' best interests to have an ample supply of tools and consumables. The tiny % of abuse that this system attracts is a very good trade off for removing an obstacle to productivity.

For the purpose of removing the constraint the WorkFace planner must be confident that the system can adequately supply tools to the workforce.

If there are specialty tools required for the task then the WorkFace Planner is responsible to order the tools and to not release the FIWP until the request has been satisfied.

The typical separation between small tools and equipment is value of the tool:

<$1500 = a small tool & >$1500 = equipment.

Tool Supply for WorkFace Planning

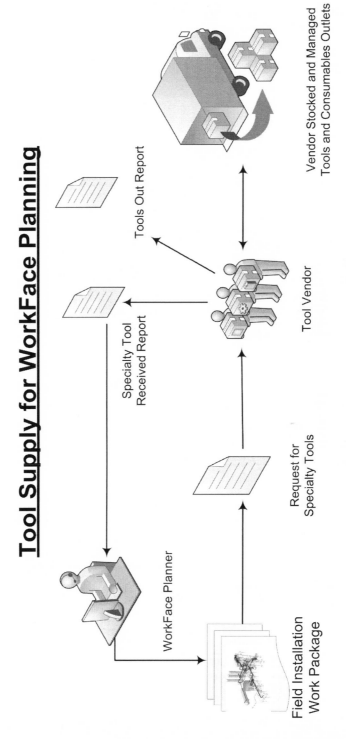

WorkFace Planner

Tool Vendor

Specialty Tool
Received Report

Tools Out Report

Vendor Stocked and Managed
Tools and Consumables Outlets

Request for
Specialty Tools

Field Installation
Work Package

9. Resources – Qualified trades with all of the appropriate site training requirements.

For a number of years in the industry this was not a concern, we just called for trades people and they would arrive by the bus load. This is still true for some sections of the industry, however more and more often now we are being constrained by the lack of qualified trades people. So it is appropriate that we understand the current reality and apply it as a constraint against FIWPs.

There are many ways to address the supply of qualified trades people and there are lots of initiatives out there right now, for our purpose we will just look at the supply that we do have against the supply that we need.

We ask the WorkFace Planners to think of trades people as a resource that we can apply to a task:

To build a concrete foundation it is going to take a lift of lumber, some forms, 15 yards of concrete, a concrete truck and driver and 45 hours of qualified trade hours. This is a common scheduling technique that draws our focus to what it will take to execute the task. In the real world of construction, the Foreman is probably thinking of things from another angle: "I have 10 guys to keep busy; I'll have to find some tasks that I can apply to this resource". Sounds like a subtle difference but one that will give us very different results.

If the resource of trades people is limited then we will have to start thinking about applying tasks to resources like the Foreman do, rather than resources to tasks like the schedule does.

If your current environment has a shortage of crane operators, when you do manage to secure some it would be a good business practice to line up a series of lifts so that you will be maximizing your progress. This is the type of information that the WorkFace Planners can use to organize their priorities around the execution of work.

The process of getting trades people to the site is normally initiated by the Superintendents and then facilitated by the labour relations team or HR department. The WorkFace Planners need to always be aware of the balance of this equation. How many and what type of trades people are we looking for and how many do we have.

The best way to accomplish this is to publish a workforce report that shows exactly how many trades people we have, which Superintendent they work for, the total number of trades people that are required and some highlights that can influence the balance. A report like this will spark conversations on the subject, which will lead to a better use of the resources that we do have.

It is a very important to note here that the WorkFace Planners need to be in the main stream of information so that these conversations take place.

For the purpose of removing this constraint from the FIWP the WorkFace Planner will take direction from the General Foreman on how much work can be released to the field. It is important for the WorkFace Planners to show in Pack Track that FIWPs cannot be released due to resource constraints if that is the case. This will show the effect of trade shortages.

One of the other issues that can become an obstacle to productive activity is the site-specific training that the trades people need to execute their normal duties. This includes operating manlifts, confined space entry, harness training, permit training and being fit tested for facemasks. We typically know before somebody gets to site that they will need to complete a series of training sessions to be effective, so we need a plan in place to address the training requirements before the candidates are brought to site. Site training that pulls the trades out of the field is a very disruptive process and should be avoided if at all possible.

10. Quality documentation – The reference documents that will govern the scope identified.

There was a time when we could just build stuff and it would be done. The world we live in now is quite different to that model and for good reason. There is great value to the customer in knowing that the product that we deliver is well built. So if we care about how well built it is, then we can care about it being fit for purpose as well. This gives us an opportunity to change our deliverable from "a plant built to spec" to "an operating business unit." The vehicle that takes us down this road is the FIWP. We use it as a process to organize and collate our Quality Control (QC) documentation so that we have a clear picture of what was built.

The interaction that is created between the work scope and the QC team is a great example of how the FIWP can work as a communication tool. The Inspection and Test Plan (ITP) is normally in place at the start of every project and it maps out the Quality Control intervention points that the contractor and owner have agreed to. The problem in the past has been to integrate this plan with the work scope, a constant challenge for the QC team. By designing a process where small well defined scopes of work must be reviewed by the QC team prior to release we have effectively created a hold point on the entire scope of work (in this case that is a good thing). The QC team can then add the documentation that is specific to the work and even some comments about preservation or handling that may not normally be communicated.

As we established WorkFace Planning teams we make a point of asking the QC people what their problems were on the last job and then look for ways to address those issues through their interaction with the FIWPs. Most problems are rooted in communication so we ask them to think of the FIWP as a personal conversation that they are having with the Foreman. "If you could talk to every Foreman just before he starts the work, what would you say?" Stick that in the FIWP.

While we like to make the lives of our QC guys more meaningful, our real target is the Foreman's effectiveness. The Foremen on most projects struggle to understand exactly what QC documents they are supposed to fill out and when. This is due to poor communication, not lack of effort. 9 out of 10 Foremen and General Foremen never get to see an ITP and yet it is their road map to completion. Strange, but true.

This process of routing every FIWP through the QC department allows the Foreman to focus on exactly what he has to do to satisfy the ITP.

The other great deliverable that this process gives us is the communication that happens after the work is complete and the documents have been filled out. The FIWP now contains the documents that we need for turnover. So part of the process is to return every FIWP to the QC team so that the original documents can be extracted and replaced with copies. The originals can then be filed by turnover system until all of the documents for that system have been collected.

To anybody who hasn't struggled to find documents to turnover a system this probably seems pretty simple. To those of us who have suffered at the hands of crappy document collection this is a "pinch me" moment.

If the FIWP comes back and the documents are incomplete then the QC rep can head straight to the Foreman and find out what happened.

Quality Control Interaction with WorkFace Planning

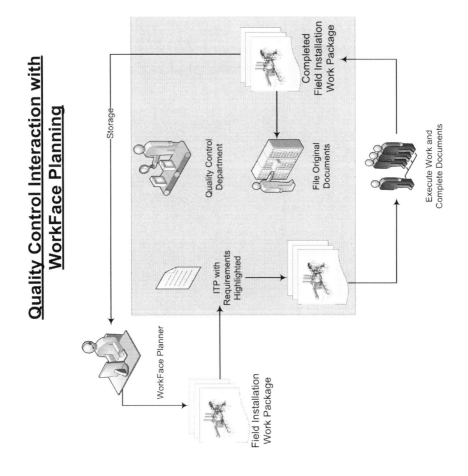

WorkFace Planner

Field Installation Work Package

Storage

ITP with Requirements Highlighted

Quality Control Department

File Original Documents

Completed Field Installation Work Package

Execute Work and Complete Documents

This is a great example of how we can improve everybody's effectiveness by developing processes that are fit for purpose.

11. Safety Planning – Information that will support the Foreman's safe application of the work.

There is a commonly held misconception about the relationship between Safety, Quality and Productivity and that is that we need to trade them off against each other:

We can have things safe and good but it won't be fast, or good and fast but it won't be safe, or fast and safe but it won't be good.

This is <u>not</u> true.

The principles that drive efficiency in any of these fields are constant.

A clear vision of what we want. (Safe, Good & Fast)

A detailed plan to get there. (Safety Manual, ITPs, FIWPs)

A process that will enable the plan. (WorkFace Planning)

We expect that the safety section of the FIWP will address all of these areas and provide the foundation that will allow the Foreman to create a culture of safety.

When we are building the process of WorkFace Planning on a specific site we encourage the Safety Department to think about their interaction with the plan in the same way that we present it to the QC group, as a communication tool. Then we ask the Safety team to read through the scope and think about what they would say to the Foreman who was about to start work. This is not their last chance to threaten him with dismemberment but a chance to add their personal summary of the dangers that may be present and the strategies that the Foreman could apply.

We have seen some very innovative safety pages that become a highlight of the FIWP. One Safety rep would always add a joke of the week in this section and then go on to provide some toolbox talks that were relevant to the work. We saw another example where the safety rep added a quiz. The end result is that this page can become very personal and it is a great way to lay the foundations that will allow the Foreman to create and maintain a culture of Safety.

The flow chart at the start of this chapter that shows the removal of constraints shows that the FIWP goes to the safety department two weeks prior to release. This is based upon the assumption that the Safety team can have the FIWPs for a period of one week. It is significant that we give the Safety team lots of time to review the FIWPs so that the task is treated as important and not urgent.

There are also a series of standard documents that should be added to every FIWP under this section. All construction sites have a toolbox talk, sign-in sheet or task preparation form that must be filled out daily. This gives the Safety team the opportunity to make sure that everybody has the appropriate documents and that any updates are distributed to the whole site.

As I mentioned in the QC section the FIWP also works well as a two-way communication tool from the Superintendents to the Safety Team by showing them where the workfronts are. If we are coming out of the ground and starting to build structures then the Safety team can adjust their overall strategy.

For the purpose of consistency the Safety Representative must sign off on the FIWP prior to its release.

12. Access to the workface – Permits and congestion from other activities/trades.

The fourth principle of productivity is "Access to the workface".

(Information, Tools, Materials, Access to the workface & Desire).

This statement covers a wide range of conditions that must all be considered when releasing work to the field. The simple way to address this is to draw on the experience of the WorkFace Planners by asking them to make a list of what could stop the crew from getting access to the workface.

The list will probably include: The issue of permits, other crews in the area, overhead lifts, road closures and even extreme weather conditions. With this list developed, the WorkFace Planners can then start to apply the principles that we use in a risk assessment: what is the likelihood of it occurring, what are the consequences and should we do something about it (mitigation strategies).

If the work is going to require a permit, the WorkFace Planner can highlight the fact and then guide the Foreman on how to obtain a permit. To find out if there will be other crews in the area the WorkFace Planner may ask the other Planners in the team or check the Four Week Look Ahead.

The result will be a one page summary of the risks as a "heads up" statement to the Foreman.

Each week we would expect to see a coordination meeting held by the General Superintendent with the Superintendents for each discipline, the Lead WorkFace Planner and a Planner from each discipline. The agenda would look at the work fronts for each discipline and see if there were any potential clashes for access. This would give the WorkFace Planners information that they can include in the FIWP that may help the Foreman execute the work. Ultimately, the General Superintendent has the authority to settle conflicts of access where the Lead WorkFace Planner cannot find resolution.

One of the things that we have been taught by the WorkFace Planners along this path of discovery is the true value of the 3D model as a communication tool. On a recent project, the WorkFace Planners successfully lobbied the Construction Manager to implement a policy where every single meeting had a live copy of the 3D model projected onto the wall. There was a joystick in the middle of the table and anybody could pick it up and use the 3D model. The WorkFace Planning software then allowed the users to define a point or clarify specific issues. It turned out to be a very powerful communication tool that increased the value of all types of meetings on Cost, Progress and even Contracts as well as all of the construction meetings.

I have heard some of the staff from that project complaining about the fact that their new projects don't have the live model in their meeting rooms. When you take a leap into the future, it is then very difficult to be comfortable in the past.

Chapter 3

The WorkFace Planners

When we first started to pull together the best practice model for WorkFace Planning with The Construction Owners Association of Alberta, (COAA) one of the obstacles that kept coming up was: Where do we get these Planners from? Our research had shown us organizations that had effective Planners and one of their key attributes was that they understood the work. This was usually because they had a trade background. However, these Planners were very hard to find so we knew that if we were going to be successful we would need a process that could create lots of good quality WorkFace Planners.

We started by creating a vision for what a perfect WorkFace Planner might look like and then had a long hard look at the current pool of potential candidates. The SAIT Fundamentals training program is the process that was developed by ASI to start candidates down the road to becoming effective WorkFace Planners.

The process for developing candidates to the point where they can build packages continues for some time after the formal training and can take between 4 and 8 weeks, dependent upon the Planner's environment, their ability to learn and the capacity of their instructors.

Potential Candidates: Working in a Union environment it was a simple task for us to advertise at the union halls for Trades people who would like to become planners. Even in an environment where we were experiencing 100% employment, we received many more applications than we needed. The obstacle that some of the candidates had to overcome was that a number of trades people don't have resumes. So we made the resume an option and then would let the candidates guide us through their history during the interview.

One interesting observation that we made was that a number of our trades people have university educations, but had chosen to become trades people in order to make a decent living.

In choosing which candidates to invite for interviews we looked for these attributes: A trade background, Supervision experience, computer literacy and some form of recent training.

When we were selecting candidates after the interviews, we tried our best to use the three C's model that Bill Hybels talks about in his book "Courageous Leadership". Character, Competency & Chemistry, in that order. The logic here is that you can develop competency but you cannot develop character, so you should choose people who have good character, not lots of experience. Then Chemistry should be low on the list because people that you don't necessarily connect with, may still be a good fit or add equilibrium to the team. Another consideration was the Team's balance between experience and youth. Too much of either one is not good.

After going through this process and building teams for 6 separate projects we did determine that the perfect WorkFace Planner was a computer geek who had 30 years of construction experience and an open mind. (We didn't manage to find anybody like this).

Key Attributes:

Trade Background

The ability to truly understand the actions that will be necessary to execute a scope of work is critical to the quality of the FIWP and the whole WorkFace Planning program. This ability usually comes from first-hand experience (and suffering) in the field. The ultimate success of WorkFace Planning will come when the General Foremen and the Foremen adopt FIWPs as their tool of choice. For this to happen, the FIWPs must survive their initial scrutiny and then prove to be beneficial. The relevance of the FIWP and its connection to reality are the key components that will make the plan work. This level of FIWP quality is most easily achieved when trades people make plans for other trades people. Dog talking to dog with no cats in the middle. This also extends to become trade specific, we should only ever see Electricians making plans for Electricians.

Test this logic by thinking about a very competent Carpenter who is assigned the task of making plans for the development of Engineering Work Packages.

Supervision Experience

When our project team first started putting together the prerequisites for a WorkFace Planner, we thought that we should have General Foremen because they understand the association between the schedule and the work. This is important and we still start most projects with this model, however we have had some very good WorkFace Planners come from the ranks of the Foremen, so you be the judge. There are lots of people out there who want to make a difference, who will gladly jump at the chance to drive us towards effective construction management.

All but two of the first 40 WorkFace planners that we hired had jobs when they applied with us. This showed us that they did not join our team because they needed the work and in some cases, they actually earned less with us than in their last position. I have always maintained that people join the ranks of supervision because they feel that they have something to contribute. When I describe the WorkFace Planning model to supervisors, the keeners are all asking how they can get involved? This supports the logic that people actually do want to be effective and they see WorkFace Planning as a tool that can help.

So the real reason that the Planners need to have Supervision Experience is because it shows that they are already trying to make projects run more effectively.

An interesting lesson learned by the Insight-wfp team: We were on our third project with 32 Planners trained before we had one quit on us (for family reasons). That is when it occurred to us that we were doing a great job of training these guys. We looked across the three projects and could not think of one Planner who was not doing a great job. We were still patting ourselves on the back when we started to talk with groups about the difference between the traditional process and WorkFace Planning. It was pointed out to us that before WorkFace Planning came along somebody was already doing the planning, they had to be, somehow projects were being completed. Those conversations led us to understand that the Foremen and General Foremen on a traditional project are doing the planning, mostly in their head or on the back of cigarette packs but it is being done. So that means that we didn't perform the training miracle that we had envisioned, all we did was to take a group of Supervisors who were already Planners by default, and give them the tools that they needed, free access to information and put nothing else on their plate, Voila! Effective WorkFace Planners.

Computer Literacy

The two core competencies that an experienced WorkFace Planner will use every day are Construction knowledge and Computer skills. We recognized these two elements very early in the development stage. Then we talked about how we could end up with WorkFace Planners that had both of these attributes and it came down to this equation:

It is easier to teach a Construction Supervisor computer skills than it is to teach a Computer Geek construction skills.

So that is why we are not too worried about what computer skills the candidates bring to the table. Over the last 10 years most of the Project Managers you know (including me) have gone from learning how to turn a computer on to where we are today, so we should be able to do the same thing with the WorkFace Planners (in much less than 10 years).

I have found that most of the candidates have a working knowledge of computers and are quick learners with the 3D model. The common language of the field is Excel and by utilizing this tool most supervisors already understand the basic elements of data management.

An interesting point here that we will spend some more time on later in the book. A key component of each WorkFace Planning Team is the 3D Model Administrator. This is typically a computer geek that has little or no construction knowledge. They are critical to the effective execution of WorkFace Planning and are a constant teaching resource for the Planners. My advice to you is to get one, pay them well and keep them close.

Some form of Recent Training

One of the issues that we face as adults is that we don't learn too good. (Like the correct use of grammar). In high school and university, our brains are predisposed to absorb information. Then we become adults and have to draw on that knowledge to make decisions. Pretty soon, we get use to our brains having output without too much input. In order to travel the road from where a supervisor is today to becoming a fully functional WorkFace Planner they need to absorb lots of input.

There is a strong argument in the medical community that suggests that the more work we put our brains through the longer they remain useful to us. Learning a second language in the second half of your life is supposed to be a good way to avoid Alzheimer's. I am not sure if learning WorkFace Planning will hold off Alzheimer's but it will certainly stretch people's understanding of the world as they know it and cause some brain pain on the way. If you can start this process with candidates that have a history of learning (recent training in anything) you will be in a much better starting position.

This was one of our lessons learned from a previous project where we taught 500 Field supervisors how to operate the 3D model. The logic that young people learn faster than old people works if the older people are not still learning. When you find a 40 year old that is taking night school for something (Chinese cooking) they are very easy to teach, especially when they have a vested interest in the outcome. The secret is to find people who have an inquiring mind.

If you sought out and purchased this book you are probably one of those types, your compatriots on this journey of discovery will probably also be inquisitive types. Luckily, for you and me, the process of WorkFace Planning seems to be fertile ground for the development of Zealots, so keep your awareness open, you'll need some friends.

Formal WorkFace Planner training.

When you have found your candidates and engaged them as WorkFace Planners the next step is to give them some formal training. You have probably started to build a vision for them that has outlined the concepts of increased productivity based upon detailed plans. A typical newbie WorkFace Planner will have caught on to some of the enthusiasm that you have shown and identified with a few of the key words that you spoke but the concepts have not yet made the transition to a concrete understanding for them.

As adults, we learn new things by attaching them to old things, so it is very important that the process of developing WorkFace Planners is structured based upon existing knowledge. The path should then be paved in logical, sequential stages that break down concepts into bite sized pieces that can become concrete knowledge through experience.

Start with the big vision, break it down into what they have to do on Monday morning and then try it, in a practice environment.

This is where formal training has its place. We have in the industry today a very well structured program that was specifically designed to help newbie WorkFace Planners get started down this road. The "Essentials" and "Fundamentals" training provided by ASI through SAIT allows WorkFace Planners the freedom to ask questions in a safe environment where the whole the room is learning and experimenting with new concepts.

For a closer look at how this training is structured please follow this link to ASI (Ascension Systems Incorporated) *www.ascensionsystemsinc.com*

The Project Team

As mentioned earlier in the chapter the ideal ratio of Planners to Craft is 1 to 50. Your logic is probably telling you right now that this is far too many or way too few. It seems that everybody has a well-formed opinion of what this ratio should be. Your own application of WorkFace Planning will give you a better understanding of the correct number. An important guiding principle here is that your objective is not to optimize the WorkFace Planners; it is to optimize the efforts of the 50 trades people. Your ability to maintain this focus will be tested many times, so remember your initial purpose was to drain the swamp not to fight the alligators; Enable productivity at the workface.

Working with the 50 to 1 ratio and based upon the assumption that you intend to apply WorkFace Planning across all disciplines, you could expect to see a Team that looks like this:

WorkFace Planning Functional Organisation
<$500 Million Project

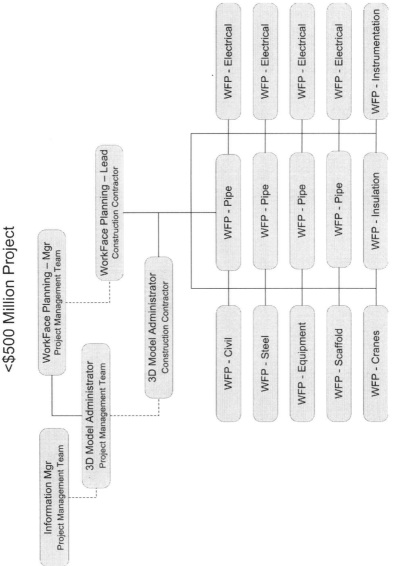

The staff positions within the Construction Contractor's organization are typically attributed to indirect costs, (i.e. not part of the direct construction costs). The other positions are members of the Project Management organization that will support the contractor's WorkFace Planning Team. It is important to note that all of these positions are dedicated to WorkFace Planning and should not be part of a shared assignment.

Each position will need a desk, high-end computer, WorkFace Planning software, high-speed internet, a phone, access to a site radio and be in the main stream of information. Don't be tempted to go cheap on this stuff, this department will make 1000 construction workers productive every day. At today's prices ($150/man-hour) this workforce will cost you $1.5 Million per day or $400 million/year (whether they get stuff done or not).

The ideal environment would see all of these WorkFace Planners arranged in cubicles, in an office complex that also houses Project Controls, Field Engineering, Quality Control, Safety and Construction Management. The complex must be as close as possible to the construction work fronts. The team will also need at least two private meeting rooms that can house up to 25 people, equipped with conference phones a high end computer, projector and screen. If you want to move into the high end of efficient construction execution (and save even more time and money) then you should equip each room with video conference equipment. This arrangement will allow the WorkFace Planning department to evolve into the natural heart of the project. Supplying information (blood) to the head, hands and eyes of the project.

Costs

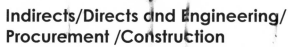

Indirects/Directs and Engineering/ Procurement /Construction

I like to think of the construction workforce as a big hairy monster that eats capital at a rate of $1.5 million/day. Luckily for us, this monster wants to be our friend and is very well trained, it can perform wonderful feats that contribute to our success and profit. However, we have to let it off the chain for this to happen. The constraints (chains) of incomplete engineering, poorly sequenced procurement and the lack of cranes and scaffolds, are stopping the monster from performing. And this is happening so that we can protect budgets that are minuscule compared to the monster's daily consumption. From a project perspective this is <u>nuts</u>.

So when we are talking about the cost of WorkFace Planning we have to think about it in the holistic context of the entire project. We are trying to reduce the Total Installed Cost (TIC), not the series of tiny budgets that we created for cost control: Cranes, scaffold, tools, IT, etc.

If you hire 1 extra person for every 50 people in the field you can expect that your total cost of construction should increase by around 2%. This will appear as an increase to your indirect costs and a decrease in your direct costs. Get ready to defend this, there are people in our organizations that hate the word "increase" and think that we should only ever "decrease".

As you read on through the book, you will also see that the application of WorkFace Planning will have a detrimental effect on the cost of Engineering and Procurement, this is the price of quality and construction productivity. The standard ratio between engineering and construction hours is 1 to 10. One hour spent engineering equals 10 hours during the construction phase. So if we can improve the quality of our engineering deliverables for the benefit of construction then the return on investment is 1000%.

Based upon the productivity statistics in our industry (40% of every worker's day is spent on direct work). At this rate, we know that we can recover the total cost WorkFace Planning if we can improve the field's productivity by only 1% (6 minutes/ day).

If we were to be so bold as to assume that WorkFace Planning could improve productivity by a further 1%, we would reduce the cost of a 500 million dollar project by a further 4.5 million dollars.

500 million x 40% (construction component) = 200 million

1% increase in productivity = 2.25% of the entire cost of a day (at a rate of 40% productive activity).

$200 million x 2.25% = $4.5 million, (for the whole project).

These figures also work in reverse, if we don't apply WorkFace Planning and then introduce an incumbence that interrupts the flow of information, tools or materials to the field then we add $4.5 million dollars to the TIC for every 6 minutes of delay.

And that is why we should be doing everything humanly possible to enable construction.

The experts in the industry are estimating that WorkFace Planning can reduce our cost of construction by 10 to 25%, this translates into a 4-10% reduction in Total Installed Cost.

Chapter 4

Summary of Basic Principles

As mentioned in the opening pages of this section the key components of WorkFace Planning are Plans and Planners (FIWPs and WorkFace Planners).These two components then facilitate the process of removing constraints, which creates a current reality for each FIWP. This gives us small packages of work that can be executed, built by planners who understand the work.

This is not how we do business now. On a traditional project, the Planners put together a plan that is a reflection of the schedule without consideration for reality and then leave it to the field supervisors to create the miracle.

WorkFace Planning demands that the work is achievable before it is released. This is probably the biggest change that WorkFace Planning introduces and the place where you will have the most resistance. The logic behind this process is that the fastest way to get the project done is to do the right things at the right time.

We already use this logic in "shutdown" environments, where the length of time that a facility is off line for maintenance is critical. Think about a shutdown where the all of the work is inside a vessel. Let's say that something happens and we are having trouble getting the manway open. We don't move our crews onto something else, we put all of our energy and focus into getting the manway open. That is how we have to think about construction, do the right thing at the right time and put all of our energy and focus into removing obstacles, so that we can stay on track. Plan the work – Work the plan

The second piece of the puzzle is the WorkFace Planners. When we first rolled out the model for WorkFace Planning this is where we thought that we would have the most trouble.

How do you create a whole new demographic of WorkFace Planners?

We put a lot of emphasis into creating a training program and identifying the pools of candidates. This helped us to minimize the issues around developing WorkFace Planners and it is now one of the areas where we are achieving consistently good results.

If you are committed to applying WorkFace Planning here are some actions for you from this first section of the book:

- Develop the model for your FIWP before the engineering starts. – Do this from the Foreman's point of view: What do you need to have to execute work?
- Make a detailed list of the constraints and then build processes that will remove them. Stay away from WAA (Wild Ass Assumptions).
- Identify where your WorkFace Planners will come from and how you will train them.
- And: read the rest of this book. We will talk about the things that you have to do to prepare the project to support WorkFace Planning (this is the really hard stuff).

Chapter 5

Resources:

As soon as you start to implement WorkFace Planning there are some obvious road blocks that appear.

In this section, we will look at the way that we will need to supply the resources of Scaffold, Construction Equipment and Material Management, if we are going to support WorkFace Planning.

The following chapters are organized into this general flow:

Why Change

Benefits

How to

Optimize

Chapter 6

Scaffold:

Identified and built fit for purpose:

Scaffold Management is one of the really big opportunities that we have for improved performance through data management. We also think of it as low hanging fruit because it is relatively easy to get on top of.

Our current systems allow a free-range process for erecting and dismantling scaffolds, which works OK, but doesn't really give us an understanding of what is going on. We know that our cost of scaffold is 25% of our direct costs and that we only budgeted for 15%.

We know that the field supervisors mean well and that the scaffolders want to do a good job. We know that the field supervisors are always experiencing delays due to the lack of the right scaffold at the right time. And we know that everybody would like to understand what is going on.

We tackled these issues on one of our early WorkFace Planning projects and achieved amazing results.

At the half way point of the project we knew exactly how many scaffolds were erected, when they were scheduled to be used and by whom, when they were expected to be torn down, how much each scaffold was costing us on a daily basis, how much each scaffold cost us to erect and a projection for what they would cost to get torn down, what the breakeven point was for leaving it up or tearing it down, how many scaffolds we needed for the average crew per week, how many scaffolders we would need in two months time, etc, etc. You get the picture: we knew what was going on and we could project costs, material and manpower requirements with a wonderful new level of accuracy.

But this is not why we did it. By far the greatest benefit was that our crews were not experiencing delays due to the absence of scaffolding. This is worth 10 times the benefit that we also got from the management of scaffold costs.

In a now famous management meeting, in the early stages of this project, a construction manager was being questioned about the higher than budgeted cost of scaffold. Armed with this scaffold data the manager explained in great detail the level of understanding that he and his team had for every single scaffold. He then went on to show the projections for scaffold utilization and the decreasing cost of scaffold erection in relationship to the direct workforce. The subject of scaffold cost was never again questioned.

And this is the third benefit: Project Management Teams and Owners want to know that their resources are being well managed.

In summary, we added some costs up front, a Programmer, a Scaffold WorkFace Planner, some computers and office space and we brought the scaffold supervisors in earlier than we would normally.

The return: an overall reduction in the cost of Scaffold to around 17% of direct work, (from 25%), the absolute minimum amount of delays due to the absence of scaffold and an increased level of confidence for the Owners and Project Management Team.

Is that an investment that you would like to make? ....
Read on.

WorkFace Planning Scaffold Request

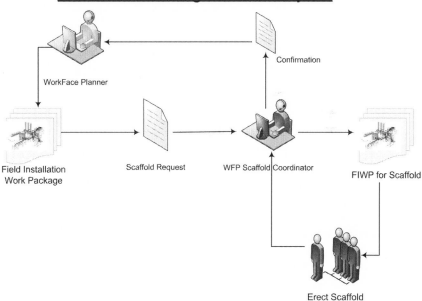

This is the basic model (from Chapter 2) for how the information flows around the WorkFace Planning scaffold management process. The piece missing is the scaffold database.

The idea of a scaffold database is not new. We have seen a variety of scaffold management processes applied with varying degrees of success. The key ingredient that was missing was the structure of WorkFace Planning.

Every data management process needs consistent and complete data to be functional.

It is of no value for the Construction Manager to report that "We have erected about half the scaffolds that we think we will need and we have spent about half of the budget".

The weak point in this argument and the weak point for most applications of Scaffold management is the lack of consistently good quality data from the field.

Traditional scaffold management systems that are aimed at managing the cost of scaffold rely on the reporting integrity of the field supervisors. The majority of Foremen will submit an official request for scaffolds well in advance. However, the data looses integrity when the occasional Foreman talks to their buddy, the scaffold foreman, and gets a scaffold built at the last possible moment. This completely undermines the process of data management and delays the construction of the planned scaffolds. Once this starts the systems quickly deteriorates into a free-range process where everybody is waiting for scaffold.

The gift that WorkFace Planning brings to the process is that all direct work must be executed from a FIWP. So we have effectively corralled all of the work and then developed a process to identify and erect scaffolds even before the Foreman has had a chance to look at the work.

This is also a wonderful two way street: The scaffolders are now working in a process that does not allow them to build scaffolds that are not in a FIWP. So they have the authority to say no when they are asked to build a scaffold without a request. They simply point across the field to the WorkFace Planners and say "Get those guys to build you a plan and we will build you a scaffold". This does not help the foremen who are trying to do some unplanned work that they have found, but it does support the cornerstone requirement that all direct work must come from a FIWP.

The ideal scaffold database for WorkFace Planning would have these features:

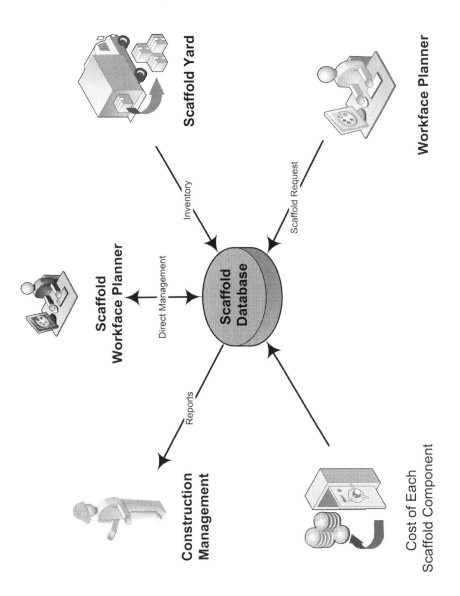

If you have a good programmer on staff, a spare scaffold expert and some time you can build a rough copy of this database yourself, or you can wait for the WorkFace Planning software to develop to this stage (2011). The important outcome is to have a system that will track the erection and dismantling of scaffold, in support of WorkFace Plans.

The database that we developed to support the application of WorkFace Planning was housed on one of the project's drives with a shortcut icon on each WorkFace Planner's desktop that opens up to an input page for the program. The input screen then asked for all of the scaffold details: FIWP number, who, when, where and for what. The program then allowed the WorkFace Planners to attach a screen shot from the 3D model. The database sorts the information and produces a response from the completed input screen. A copy of the request is printed by the WorkFace Planner and then added to the hard copy FIWP.

That is the total effort needed from the WorkFace Planner. Order it and then attach a copy of the request to the FIWP. The Scaffold WorkFace Planning Coordinator now has the responsibility to deliver upon the request.

As the WorkFace Planning software evolves in this direction we can expect to see a program that interacts with the 3D model and actually shows the scaffold in the model, (but we have to learn how to run before we can fly).

The scaffold database collects requests and stores them in a summary format page for that day. The Scaffold WorkFace Planner develops each request into a package of work and then assembles a series of "builds" into a Scaffold FIWP. This should produce a Scaffold FIWP that is one week in duration with scaffolds that are required two weeks from the date of issue.

Imagine that....!! Scaffolds built before we need them and two weeks' notice for the scaffolders.

This is LaLa land for our seasoned constructors, but it is true. We have done this and seen the results.

It is important to remember that this is our primary focus: Get scaffolds built fit for purpose and before we need them. Again: Fit for purpose, before we need them: Fit for purpose, before we need them: Fit for purpose, before we need them.

Once we have this satisfied then there are some golden scaffold management opportunities that present themselves.

The information that the database now needs is the erection details: Who did it, when did it go up, a list of components that were used and how long did it take. This information can be collected by utilizing a travel sheet in the scaffold FIWP, that has all of the scaffold information from the request and then asks the foreman to fill in quantities beside a checklist.

The check sheet comes back to the Scaffold WorkFace Planner or a data entry clerk and the information is entered into the Database against the scaffold Id number. This also automates a response to the WorkFace Planner to confirm that the scaffold was erected.

Cost:

Now the scaffold database can calculate the components used against the rental cost for each one and produce a per day/week/month rental cost for the scaffold. The database can also be populated with the hours for erection. This allows the database to calculate the cost of labour to erect and the projected cost for teardown (40% of the build).

These two figures: rental cost & labour for the teardown & erection will give us a breakeven point for the decision on whether to tear it down after its initial use or how long we can leave it up for a future use and still be economical.

Inventory:

The main scaffold yard (offsite but close) should have a computer terminal that is linked to the scaffold database, (This could be through the internet). The computer at the yard would record shipments coming in and inventory that is being checked out by the Scaffold Foremen. The scaffold database would then draw quantities from onsite materials to build scaffolds and show the remaining stock, onsite and offsite. This information can easily be summarized and presented to the Scaffold Superintendent for the purpose of ordering materials. It would also show at a glance the current stock levels, which could be very useful information for the scaffolders who are searching for components.

This scenario is only made possible by the integrity of data. If the yard is diligent about maintaining stock levels and the information is kept up to date then it becomes reliable.

Statistics and projections:

This is one of the other benefits that come from complete data management. We now know that the average 500 work-hour FIWP will require 2.2 scaffolds that have 110 square feet decks and stand 3.2 meters tall, because we tracked them. So anybody who gets serious about managing scaffold will also know how long the average scaffold takes to build. Then you take the estimate from the WorkFace Planners that tells you how many plans they will build and when, multiplied by your scaffold stats and viola, you have a very good scaffold plan and estimate. This information will give you the labour and materials required on a schedule, which can then lead to early material requisitions, right sized lay down yards and lunchroom/washroom facilities in the right location.

Now imagine developing projections based upon the information that you have stored in your database from your last 5 projects.

Accountability:

One of the subtle changes that this systems implements is the shift in accountability from: the Trade Foremen - to the Scaffolders. Primarily the shift is about who is responsible to get it built, but as important is the end result and whether the scaffold is fit for purpose.

Fit for purpose has many angles: the first is: will it work for the guy that ordered it, then will it work for the next trade that may want one in the same spot, then could the request have been satisfied as part of multipurpose larger scaffold.

These questions are most often not addressed because the current system places the accountability of "fit for purpose" in the hands of the requestor. The Foremen ordering the scaffold makes sure that it works for them without any consideration for the other trades. (as they should).

By giving the Scaffold management lots of notice and the accountability to satisfy the requests, we are also empowering them to be efficient. Ultimately, they now have the authority to build scaffolds as they see fit, so long as they satisfy the requests. This may not seem like a big change but speak to any Scaffold Foreman and they will tell you that they know where to build scaffolds and which trades will need them long before the requests come in. So if we tap this expertise and tell the scaffolders to sit down at the start of a project and think about where we could build long term multi use scaffolds we could have a big portion of the requests satisfied long before they are submitted. For the scaffolds that are built for individual requests the scaffolders can calculate the cost of leaving them in place rather than tearing them down and rebuilding them if they anticipate that another trade will need a scaffold in the same place.

It is a great example of the benefits that we can reap when we drive the decision making into the hands of the people who understand the questions.

Scaffold Management Optimized.

The model that I talked about on the previous pages is the base model that you must have in place if you want to have a functional application of Scaffold management that can support WorkFace Planning. In the following passages, we are going to look at the advanced version that you might want to look at after a project or two.

(Please remember that you don't learn how to fly in fighter jets).

The first place to start is the integration between the 3D model and Scaffold management. I know that there are products on the market right now that allow you to see your scaffold in the model. This is pretty cool when you are trying to explain what it is and where it goes, however the model quickly gets clogged with scaffold. I expect that the software developers will work out a way to visually represent intelligent scaffolds in a manner that will not clutter the model sometime soon. The progressive stage towards this outcome may be to have the database store 3D model coordinates that can zoom the user to a point in the model. I have seen both of these functions demonstrated so they are possible now.

Later in the book I will be talking about the total management of information and a position within the WorkFace Planning team for a 3D model administrator. The ideal candidate for this position would also have the capacity to develop databases in Access or SQL. This will allow you to develop a scaffold database, like I did, where you identify the inputs/outputs and basic functionality and the database guy does the rest. It is a slow process that will take a couple of months to initiate and then constant management for the life of the project. Please keep in mind that in the world of mega projects, programmers are cheaper than flag people (the guy with the stop/slow sign) so get a good one and pay them well.

If you are going to get serious about managing scaffold data and you want a database that you can use for the next 10 projects then consider developing a database that it is ISO 15926 compatible and xml based. (Check out ISO 15926 on Google).

I have always advocated that our Project organizations should try to stay out of the software business. This is because one of the cornerstones of business is that "You make money when you do what you are good at". We are not particularly good software developers but we are good at managing projects and building stuff. The problem is that we need to be able to manage scaffold information so if you cannot find a product on the market you have to build one yourself. You could spend the time and effort to develop a detailed scope of functionality and then engage a software developer to build you the right tool, but like most other IT projects, it is very difficult to develop a detailed scope.

So software developers take note: The industry needs a scaffold management software that will support the application of WorkFace Planning (and interact with the 3D model).

We just want the tools that allow us to do the work.

Scaffolds to Work Stations;

This is a developing field that has lots of potential to create productive activity.

Imagine yourself as a welder who is told to weld two pieces of pipe together 40 feet off the ground in a pipe rack. If you are lucky enough to work in an environment where WorkFace

Planning is being applied then you probably already have a scaffold built and it is in the right place. Now all you have to do is to take your cables, grinders, power supply, fire extinguisher, water bottle up there and you can start work.

What if all that stuff came with every scaffold?

This is not a new idea and most good field supervisors will have apprentices running around preparing these scaffolds to be workstations, but we can also do it when the scaffold is being built.

There is at least one scaffold supply company that has a product line that is designed to do exactly this.

They have a power board that clamps onto the scaffold and connects a high amp power supply directly from the temporary power grid. Most of the facilities that will be needed are available as options that come with scaffold mounts, including the spark containment. To make use of this function we would need to expand the scaffold request form that the WorkFace Planners use to include "Scaffold Options": Power, lights, spark containment, fire extinguisher, grinder stands, water jug etc. There could also be some standard items that come with every scaffold: a garbage can and a bucket & rope for moving tools etc. The idea is that we could request and build workstations not just scaffolds. When the industry finally wakes up and moves to lunchbox welders then the problem of welding cables will also disappear overnight.

This opens up a whole new dimension of planning. The logic of temporary power being connected to work stations allows us to think about designing a temporary power grid even before we break ground. Then we can develop the WorkFace Planning department so that the supply of temporary power is linked to the FIWP development plan....and so on.

(Jet fighter stuff)

Chapter 7

Construction Equipment Management

For the purpose of this book we are looking at the management of cranes that come <u>after</u> the heavy lift program. The heavy lift program on any one of these mega projects sees several hundred vessels and modules escorted over the highway and set in place before the work of connecting them begins. The activity is very dominant and by its nature demands a lot of planning and coordination. The process of allocating construction equipment to lift and move these components receives a high level of attention and is normally very well planned, so there is less need for WorkFace Planning here.

For the remainder of the project:

The management of cranes, manlifts, welders, heaters and pumps is an area where we don't spend a lot of time during the preparation phase of a project. It is normally assigned a budget based upon an analogous estimate from experience and historic figures. Then a supplier is chosen based upon rates and reputation. The day-to-day management of equipment is assumed to be a negotiation that will take place between the manager (who has the budget) and the field supervisors who will be screaming for equipment.

As a component of the Field Installation Work Package (FIWP) construction equipment is critical. You cannot make a lift without the right piece of lifting equipment and you cannot keep a concrete pour warm without heaters. For the purpose of planning this allows us to view construction equipment as a resource that must be allocated to the package prior to the work being released (as described earlier in the chapter on constraint removal). No equipment = No work.

The existing problem is that the equipment budget is the dominant piece of the puzzle. The logic is that we can minimize the overall cost of a project by minimizing the cost of the individual departments. This not true, and it leads us to situations where you find that a small component of the overall cost (the construction equipment) is having a negative effect on the mother-load of the project's costs (labour). 10 cents saved on equipment is 10 dollars blown on labour. For WorkFace Planning to be effective we need to see construction equipment as a constraint on each FIWP that must be removed.

Much like the opportunities that we have in scaffold management, there exists lots of room in our equipment management processes for improvement. The key is to start with our objectives in the right sequence:

- The first goal of equipment management must be to satisfy the needs of construction through FIWPs.
- The second goal can then be to optimize the management of equipment costs.

During our research on WorkFace Planning we saw a couple of good examples of equipment management; it was usually sparked by one individual who "had a dream". The basic principles are the same for any commodity management process: What are we trying to achieve? What resources do we have? And a simple process for tracking demand, supply and cost.

The best proof that this is beneficial came to us as a negative trend. It was a situation where there were several silos of similar construction on one project and only one of them was managing equipment by design. The other six silos where asking for more funds for their equipment budgets, while the one with the equipment management process was trying to give back $750K, because they didn't need it.

While the equipment utilization rates in this one area were 25% higher than the other silos, by far the greatest benefit was that the work crews in this area reported very few instances of delays created by lack of equipment. This has 10 times the value of the costs saved on equipment rental. (10 trades people waiting / one crane and operator)

The application of WorkFace Planning requires that we can allocate resources to work prior to its release. This means that we need to identify the cranes , manlifts, pumps and heaters etc that are needed to do the work and then assign specific units to the work prior to release. This gives us the base model for the management of equipment.

- Identify the requirements
- Know and track what you have
- Work to satisfy the need of the construction work fronts.

WorkFace Planning Equipment Request

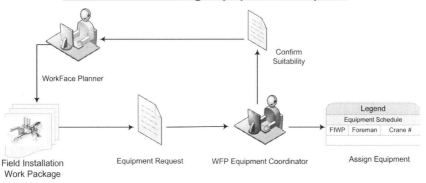

This flow chart (from Chapter 2) represents the model that we developed for the removal of construction equipment constraints.

The component that makes this work is the equipment management database. It is based upon the same principles as the scaffold database.

At this time, I have still not developed one of these databases completely. I have seen data management processes applied and I have structured equipment management systems to satisfy WorkFace Planning, but not with the same level of sophistication that we have with the scaffold management database.

So here is the outline of what it will look like when I get the opportunity:

The WorkFace Planners will have an icon on their desktop that is linked to the Equipment Management Database. A double click will bring up an equipment request form that requires all of the pertinent information from drop down menus along with the FIWP number and some 3D snapshots from the model, with coordinates. The WorkFace Planner will copy the scope of work onto the request and then make a recommendation for the size and style of the equipment needed to support the work. This may include some explanation:

"The work is only 40'off the ground but we will need a 125' manlift to get over the excavation which is at the base of this work front".

The WorkFace Planner will complete the request and the database will auto generate a request number that incorporates the FIWP number. The WorkFace Planner will print a copy of the request and add it to the FIWP.

The Equipment Management Database will summarize the requests for that day and add them to the master spreadsheet that shows the WorkFace Planning Equipment Coordinator a summary of all requests.

The coordinator then verifies that the equipment requested is fit for purpose and then allows the database to add the equipment requests to the equipment schedule. This schedule will show the allocation of each unique piece of equipment for each quarter of each day.

Crane Allocation Schedule

Week starting May 25th 2009

Period	Mon 1	Mon 2	Mon 3	Mon 4	Tue 1	Tue 2	Tue 3	Tue 4	Wed 1	Wed 2	Wed 3	Wed 4	Thu 1	Thu 2	Thu 3	Thu 4	Fri 1	Fri 2	Fri 3	Fri 4	Sat 1	Sat 2	Sat 3	Sat 4	Sun 1	Sun 2	Sun 3	Sun 4
Unit #																												
65-123	P1	P1	P1	P1	P1	P1	B3	B3	P1	P1	B3	B3	P1	P1	P1	P1	P1	P1	B3	B3	P1	P1			P1	P1	B1	B1
65-124			B3	B3	C2	C2	C2	C2	C2	C2	C2	C2			B3	B3					C2	C2			C2	C2	C1	C1
75-126	B1	B1	B1	B1	B1	B1	B1	B1	B1	B1			B1	B1			B1			C1	B1				B1	B1	B1	B1
75-130	P2	P2	P2	P2	C1	C1	P2	P2	P2	P2	C1	C1	C1	C1	P2	P2	P2	P2	C1	C1	P2	P2	P2	P2	P2	P2	P2	P2
80-134	Yard	Yard	Yard	Yard	Yard	Yard	Yard	Yard	Yard	Yard	Yard	Yard	Yard	Yard	Yard	Yard	Yard	Yard	Yard	Yard	Yard	Yard	Yard	Yard	Yard	Yard	Yard	Yard
125-168	E1	E1	E1	E1			I1	I1	I1	I1	I1	I1	E1	E1	I1	I1	I1	I1			I1	I1	I1	I1	E1	E1	E1	E1
225-174	P3	P3	P3	P3	P3	P3	P3	P3						E1	E1	E1	P3	P3	P3	P3					P3	P3	P3	P3
225-175	B2	B2	I2	I2	B2	B2	I2	I2	B2	B2	I2	I2			I2	I2			I2	I2			I2	I2	B2	B2	I2	I2
225-176	I3	I3	I3	I3	I3	I3	I3	I3	I3	I3	I3	I3	I3	I3	I3	I3	I3	I3	I3	I3	I3	I3	I3	I3	I3	I3	I3	I3

Crews

Ironworkers	1	2	3
Carpenters	1	2	
Boilermakers	1	2	3
Pipefitters	1	2	
Electricians	1		

In this example, the cranes have been identified as unit numbers: 65-123 is a 65 ton crane, unit number 123, with allocation to crews (P1 & B4) by the ¼ of a day. The allocation could also be by FIWP. We would expect that the database would also display another level of details with a mouse click on the square. The database would break out cranes from manlifts and smaller equipment.

The WorkFace Planning Equipment Coordinator would factor in geographical and known constraints and then develop a schedule for each unit for each week. The reality of the field would also need to be factored in, and we should probably start the project with 30% more equipment than we have scheduled. This will cover breakdowns, absenteeism and allocation overruns. Over the course of the project this could be decreased as the FIWP estimates become more accurate.

The tough part will be to keep the system's focus on support for FIWPs, not minimal equipment costs.

The Equipment management team (Foremen, General Foremen and Superintendents) will manage the day-to-day allocation of cranes and Equipment Operators to satisfy the Equipment schedule.

Once this system is in place, the WorkFace Planning Equipment Coordinator will establish a validation team (typically Labourers) who will track the utilization of each piece of equipment through industrial engineering observations. Each unit number is observed 4 times a day for being used /or not and then the results are tracked and graphed. A simple database will allow the information to be sliced by unit number, user, user group or by days of the week. The low number of observations taken across a one-week period would not give a definitive utilization figure but can be used to flag potential allocation problems. The cumulative results over a month by month graph could show trends in overall utilization and be used for future estimating.

Cost: If we are tracking the allocation and utilization of a piece of equipment we could also easily add the daily cost to the database.

The initial benefit is that we can track a unit's cost per month and summarize the costs for each Superintendent, General Foreman or Foreman. This is sometimes an eye opener for the Supervisors and can help them to understand where some of the project's costs are coming from.

I have seen simple examples of this used on recent projects where the hourly cost of a piece of equipment is displayed on a large sticker that reads: "This piece of equipment costs the project $125/an hour".

The database could also be structured to help the WorkFace Planning Equipment Coordinator optimize equipment utilization rates. This would work well when we have contracts that stipulate cost up to a certain number of hours per month and then a discounted rate for hours above that number. Once programmed the database could then alert the Coordinator when a piece of equipment is approaching optimal utilization.

This diagram shows the basic requirements for a functional equipment management database.

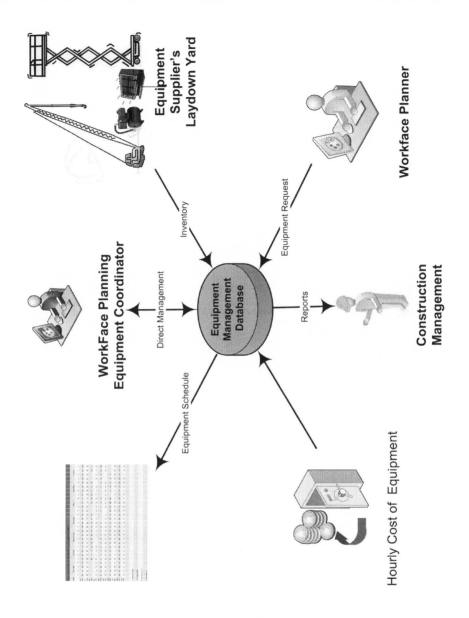

Equipment Supplier's Laydown Yard

Workface Planner

Inventory

Equipment Request

WorkFace Planning Equipment Coordinator

Direct Management

Equipment Management Database

Reports

Construction Management

Equipment Schedule

Hourly Cost of Equipment

The WorkFace Planning Equipment Coordinator will draw weekly reports that show each supervisor a list of the equipment rented under their name and ask them to confirm that they still have the units. This validation process will improve the accuracy of the billing process and allows the contract administrators a way to confirm accuracy and facilitate payments. Management reports from the database will be rolled up to a level- 3 standard where the total cost of equipment will be reported against level-3 activities in the schedule (more about this later). The WorkFace Planners will receive utilization reports to improve the accuracy of their future FIWPs.

Much like the scaffold management process, this process moves the accountability for satisfying the FIWP resource requirements onto the WorkFace Planning Equipment Coordinator. This will empower the equipment management experts to make decisions with the focus of satisfying the project schedule. For this process to be successful the WorkFace Planning Equipment Coordinator will have to have strong support from the Project Management Team, be enrolled in the overall vision of WorkFace Planning and have the authority to drive the system's goal:

Satisfy the needs of the construction through FIWPs.

Equipment Management Optimized.

The basic step for equipment management is to develop a system that satisfies the needs of FIWPs. This can be achieved by following the process laid out in the last couple of pages. This next section will be for your Team's next project after you have mastered the process of satisfying work fronts.

The development of a database that itemizes each piece of equipment by unit number takes us naturally into some other areas that we have typically struggled with: maintenance and fueling.

Maintenance:

The reliability of equipment is a major contributor or detractor of productive activity so a maintenance program that reduces equipment down time is money well spent. By utilizing the database to manage the maintenance program we can automate the application and reduce the risks that a piece of equipment may get missed. Once we have a unit number entered and we have programmed the frequency and type of maintenance that should be performed we can then have the validation team collect hours from the hour meters once a week and enter the information against the unit number. This information could be drawn weekly by the maintenance team from a web based log in to the database, who then add their maintenance records for the last week.

An advanced version of this process is already being applied by the aeronautics industry where they utilize RFID (Radio Frequency Identification) tags and a hand held scanner. The maintenance guys simply scan an object's RFID tag and the full history for that unit comes up on their scanner. They perform any maintenance that is required and then enter the information onto their scanning unit. A download at the end of the day updates the database online.

For our application, we could advance to the point where our validation team scans the units and enters the hours from the hour meter. The maintenance team then builds a maintenance schedule for the coming week and book equipment out on the master schedule for certain time slots.

Fueling:

Imagine yourself as a fuel truck driver working night shift, it is -20 degrees and you know that there was a series of critical concrete pours that happened during the day and your job is to find the heaters that are keeping the hoardings warm overnight and add fuel. The site is covered with over 100 scaffold hoardings (tents) and you have to figure out which ones have heaters.

This might sound like a fun game but it is not.

We often have significant incidents that have a major impact on productivity due to equipment running out of fuel. It sounds like a simple issue to fix but it is wildly complicated. The fuel truck driver needs to figure out where every piece of equipment is and every day all of the locations change, equipment is removed, new units are added and new areas of construction are started.

The solution utilized by most sites is that the fueling is conducted during the day. This gives the fuel truck driver access to the construction workforce and lots of important information. The problem is that most equipment must be shut down during the fueling process so we are reducing the utilization window for each unit. This is not much for any one unit on any given day but the cumulative effect is an erosion on the project's productivity.

The Solution: There is at least one equipment supply company in Canada that is experimenting with GPS locators to address this issue. The logic is that each unit would have a GPS locator, which would allow the fuel truck driver, the maintenance guys and the equipment management team to always know the exact location of every piece of equipment.

This opens up a series of other opportunities like automating the database to show the GPS locations of equipment and getting maintenance performed on nightshift.

One of the other sad realities for equipment managers is that we often lose equipment on our own sites. Most of the time this is not through pilferage, but is through 'borrowing", where a neighboring silo borrows a manlift or heater and then forgets to return it. The unit then sits there on rent and not utilized until it is found or returned. GPS locating and tracking will resolve this issue and allow the Equipment management Team to facilitate "borrowing".

The problem for us right now is that GPS management is one-step too far for Construction Management at this point so we are waiting for the Equipment suppliers to step up and drive this as an extension of their competitive advantage.

<u>Take note equipment suppliers</u> : what we really want is a total equipment management package that does all of the stuff that I just talked about. We don't want to be in the equipment management business, we want you to manage it, based upon our needs. Right now we have to go there because somebody has to.

Data management:

I believe that the future of data management for construction projects will be web based. There are many examples of this already in application and they could satisfy our needs very well. Here is how it may look for Equipment Management:

The database that I just described is online and all of the projects stakeholders have access to it through access rights and passwords. The project managers have a dashboard icon on their desktop that shows them equipment, scaffold, labour, productivity, and progress all rolled up to match <u>level-3</u> activities on the project schedule. The dashboard draws live information from the online databases and shows all of the project stakeholders their view of the project. A double click on any figure expands the information to the next level of detail.

The Construction Manager looks at the dashboard on Monday and sees that the equipment costs in one area seems high so he double clicks on the figure and a graph pops up to show the trend over the last 3 months. The Manager right clicks on the graph and selects "Add Labour". Now he sees that the number of trades people has increased and also notes that the civil trades are decreasing and being replaced with mechanical trades. Knowing that the mechanical trades typically use more equipment this justifies the increased amount of equipment. The manager makes a note to confirm this balance with the WorkFace Planning Equipment Coordinator at their weekly meeting.

This scenario is not too farfetched. It will probably be 5 – 10 years before we see Project Management at this level but we have the technical capacity to do this today. We will need a constant supply of reliable information, software programs that are fit for purpose and some cultural changes that allow us to manage construction based upon WorkFace Planning.

For now try to keep your head out of these clouds and just get good at applying equipment to FIWPs. This is the path that will lead us to Total Project Management.

Summary of Construction Equipment:

The application of WorkFace Planning and Field Installation Work Packages gives us an opportunity to identify Construction Equipment as a resource. This leads us to a scenario where the equipment managers can see a direct link between the ways that they supply equipment and how much progress we earn against the project schedule.

This is not how we do business now. Equipment management teams see themselves as independent to the process of satisfying the project schedule. Their role is to satisfy the budget produced in the estimate. This has led us to where we are now: Choke Management.

Choke Management: Restrict the supply of a commodity until the screams from the end users get to a predetermined level (Sometime before they stop breathing) and then back off the choke a little bit. The screaming slows down, you have reduced the supply of the commodity and it looks like you are managing supply against demand.

(You have 5 cranes in the field, take two of them away and then return one).

This is how we do business now and it does not fit with WorkFace Planning, we still need all 5 cranes.

So the first thing that you will need to make this work is the authority to change this, and that comes from a project wide commitment to apply WorkFace Planning. Then you can go ahead and build a database that will identify the needs, (from FIWPs) and then allocate resources against those needs. (Demand and Supply).

Chapter 8

Material Management for WorkFace Planning:

The ineffective management and supply of materials is a common issue on projects that experience poor productivity. Most of the material teams that we researched were staffed by very competent individuals who tried very hard to satisfy the needs of construction and yet they still achieved poor results. Their common problem was that the information they used was incomplete and the systems used to manage the data were misaligned with the needs of construction.

So this is good news, it tells us that the people in the system want better results (change) and that the solution is in the design of processes and data management.

When I talked about constraints earlier in the book I made this statement:

"Assume that the onsite material management group can tell us what they have and what they don't have".

This is a very big statement and is the essence of how material management will have to function if it is going to be structured to support WorkFace Planning.

This statement suggests these assumptions:

"What they don't have": This implies that the onsite material management team has a full list of what it will take to build the project, every nut and bolt. If that was to be true then where did they get it from and how is the information organized. When a widget comes through the gate how does the receiver know what it is and which item gets checked off the list?

"What they have": This statement implies that the Material Management Team have started with a list of what to expect and what to call it, then received widgets against the list and are now in a position to communicate what they have.

"The onsite Material Management group can tell us": This portion of the statement suggests that the material management team can communicate with Construction and also implies that they can tell us what they have before we ask them. For this to be true we need to have a common language and a method that effectively communicates.

This is not the current reality.

Many a project manager has looked at me with slacken jaw when I show them that their Material Management Team don't know what they need, don't know what they have and want us to describe widgets so that they can look for them.

Sounds bad, hang on it gets worse:

On a recent 1.5 billion dollar mega project we found a material management system that was in total chaos. Their system was exactly as I just described. They had been told by procurement that they had everything, they just didn't have a list of what "everything" was or where it had been put down. The solution that they applied was to put more people in the material yard and try to identify material that may develop into work fronts. (Sound familiar). The construction site had about 500 pipefitters and welders, the material yard had 46 pipefitters. Our average total burdened cost for pipefitters is around $150/hour. So for arguments sake we drop this rate to $100/hour and it works out to $46,000/day, $230,000/week or about a million dollars a month to try to find stuff. And that is just for the Pipefitters.

Sounds bad, hang on it gets worse:

The 46 pipefitters were not finding enough work fronts to keep the 500 pipefitters and welders busy so now they had a $500,000/day monster that was eating project capital and not getting much work done. (Remember this is just one trade).

Net result: $500,000/day capital eating monster is sleeping on the job because it has not much to do.

This is not a fairytale with a bad storyline. I am not making this stuff up. This is common and is the result of Project WAA (Wild Ass Assumptions).

A year after I found this situation, I had lunch with one of the survivors and he said: "I now know that I don't know much, but I'm sure that I never want to do that again". I am happy to report that this manager and his organization are now WorkFace Planning Zealots.

For those of you who are crying while you are reading this, fear not we have glimpsed greatness and it is achievable. For those of you who don't know what I am talking about: Your life is too short to make all of the mistakes that you need to, so figure out how to learn from the mistakes of others.

So now lets look at what a WorkFace Planning material management system could look like and what benefit we could draw from that.

Material management is really data management, look at the way Wal-mart manages stock levels: An item is scanned at the checkout and this produces an electronic request for a replacement directly to the manufacturer. The software system does not care what it was or how it is going to be manufactured or shipped, it just wants a replacement. (data management).

There is also a solid argument that the difference in the battle of Brittan during WW2 was that the British could replace planes faster than the Germans could. This was due to a communication network that channeled the right resources to the right factories (data management).

The most effective material management systems that we have seen in the construction industry utilize the 3D model as the major source for all project information. The model produces a list of components and this is used to populate the material management database. The database then receives and issues components based upon their model name. This is made possible by the fabricators being contractually bound to identify components by their model name and then Construction using the same name to order materials. This creates a system where everybody speaks the same language: the project's Work Breakdown Structure. (Lots more to come on the WBS).

The results are systems that disappear from view. Material problems are very small and often insignificant. The management of materials in these systems operates a lot like the sun and the moon. They do their thing without much effort from us. And a lot like sunshine on a warm day it is easy to take an effective material management process for granted. It took a lot of effort and design to make the world turn and it takes a lot of effort and design to create a material management system that supports WorkFace Planning.

There are many different material management databases on the market. A lot of them have all of the features that we need to support WorkFace Planning. The key is to have them configured so that the inputs-processes-outputs = what we want.

So let's start with that:

The application of WorkFace Planning is expected to produce Field Installation Work Packages that have less than 1000 hours of work. At a point 4 weeks prior to the scheduled execution of the work the WorkFace Planner must confirm that all of the material required for this one FIWP is available. The FIWPs will be a collection of complete drawings that make up the work. At this point the WorkFace Planner will ask the WorkFace Planning Material Coordinator "Can you supply the BOM (Bill of Materials) for these drawings". The WFP Material Coordinator will answer Yes or No with a % complete. The WorkFace Planner will then either move the FIWP into the four week look ahead and order the material or if the answer is No move the FIWP back into the queue and wait until the answer is Yes.

Material Management to Support WorkFace Planning

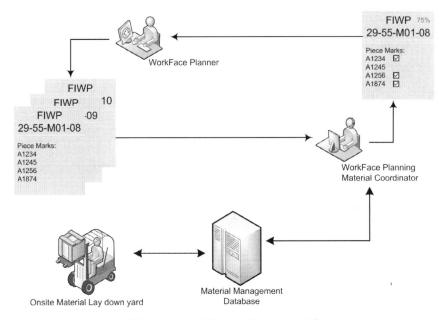

(Flow chart from Chapter 2)

Given the benefit of foresight we can predict that the WorkFace Planning Material Coordinator is going to receive thousands of these requests on any typical project, so we can prepare them to be ready with the answers or even to answer the questions before they are asked.

The following flow chart shows the series of events that will need to take place in order to facilitate this application. The numbered passages that follow the flow chart correspond to the numbered nodes in the chart.

Material Management Cycle

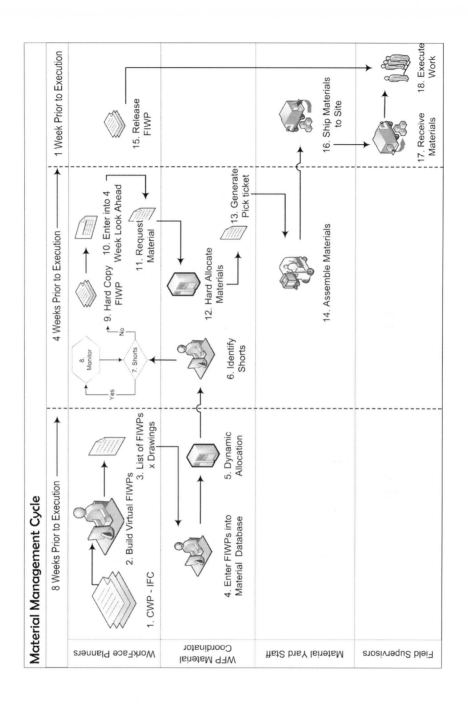

8 Weeks Prior to Execution

1. CWP - IFC
2. Build Virtual FIWPs
3. List of FIWPs x Drawings
4. Enter FIWPs into Material Database
5. Dynamic Allocation

4 Weeks Prior to Execution

6. Identify Shorts
7. Shorts
8. Monitor
9. Hard Copy FIWP
10. Enter into 4 Week Look Ahead
11. Request Material
12. Hard Allocate Materials
13. Generate Pick ticket
14. Assemble Materials

1 Week Prior to Execution

15. Release FIWP
16. Ship Materials to Site
17. Receive Materials
18. Execute Work

Workface Planners

WFP Material Coordinator

Material Yard Staff

Field Supervisors

1. A Construction Work Package (CWP) will be a single discipline and less than 40,000 hours of work (more on this later) and will be considered Issued for Construction (IFC) when all of the drawings have been entered into the document control database. At this point the CWP will be issued to the fabricator and to construction.
2. The CWP dissection process detailed in the Basic Principles section of the book will be applied by the WorkFace Planners and the Construction team. This will produce FIWPs in the WorkFace Planning software and the 3D model. (Virtual FIWPs).
3. The WorkFace Planners will produce an electronic list (in excel) that shows all of the FIWPs for this CWP and the drawings that are allocated to each one. The sum of the drawings will equal an entire CWP. The list will then be submitted to the WorkFace Planning Material Coordinator.
4. The WorkFace Planning Material Coordinator will group the drawings in the material management database and name the groups after the FIWPs. The FIWPs will be prioritized based upon the proposed execution dates supplied by the WorkFace Planners.
5. The bulk material in the database will then be dynamically allocated against the FIWPs (based upon priorities) to show their % complete (received). This process is updated daily as materials are received.
6. The WorkFace Planning Material Coordinator will produce a weekly report that shows each FIWP and the % received for tagged items and bulks. Each FIWP would also have a list of the materials that have not been received (shorts). This allows the WorkFace Planners to override the 100% material rule if the missing components are inconsequential.
7. The WorkFace Planners look at the FIWPs each week and determine which ones have satisfied their material requirements.
8. The FIWPs that have missing material can be monitored daily as new materials are received. This is made possible by the WorkFace Planners having read access to the database (more on this later).

9. When the material constraints have been satisfied the WorkFace Planners will develop the Virtual FIWP into a hard copy so that it can start the process of having other constraints removed.

10. As the hard copy FIWP is produced, the General Foreman then has the option to add it to the four-week look ahead, with consideration for prerequisite FIWPs.

11. When the FIWP is added to the Four-week look ahead the WorkFace Planners will submit a material request to the WorkFace Planning Material Coordinator for the FIWP materials. The request will not detail the materials, it will only request the materials associated with the FIWP. The material management database already has a complete list of the materials for each FIWP.

12. The WorkFace Planning Material Coordinator will then hard allocate the bulk materials in the database. This removes the material from stock and allocates it against the FIWP.

13. The WorkFace Planning Material Coordinator then draws a detailed list of the materials from the database and issues a pick ticket to the warehouse, complete with the FIWP number and the Required On Site (ROS) date.

14. The warehouse staff assembles the materials and label the shipment with the ROS date, site delivery coordinates, FIWP number and contact information for the onsite Foreman.

15. The WorkFace Planners progress the FIWP through the constraint removal process over the next three weeks and then release the unconstrained FIWP to the General Foreman in the week prior to the scheduled execution. The General Foreman then issues the FIWP to the Foreman prior to the ROS date for the material.

16. The Warehouse staff ship the FIWP materials to site on the ROS date.

17. The foreman or assigned material receiver signs for the materials and checks the components against the FIWP material list.

18. The Foreman and crew execute the work.

In order to support this process the material management database would need to have these features:

- Have the same list of electronic drawings as the IFC 3D model with no extra dots, dashes or numbers.
- Allow the WorkFace Planning Material Coordinator to develop Parent- child relationships between groups of drawings and FIWP numbers.
- Be able to separate fabricated materials from bulks (field) for each drawing.
- Be able to electronically extract BOMs from the drawings and develop a list of materials required for each FIWP.
- Be able to prioritize FIWPs for material allocation based upon scheduled execution dates.
- Be able to manage bulk materials and tagged items separately.
  - Tagged items may include: Spools, steel piece marks, instruments, electrical equipment etc.
  - Bulks are cable, tray, bolts nuts, gaskets, loose fittings etc.
- Be able to receive these types of unique materials directly against the items identified in the database.
- Be able to roll up received items against the FIWPs.
- Produce a % complete (received) report for each FIWP on bulks and tags
- Be able to produce a list of shorts for each FIWP, (tagged items and bulks).

And:

Dynamic Allocation

> *An electronic function of material management databases, which identifies bulk material availability and compares it with the FIWP material requirements. The allocations are prioritized by the scheduled execution dates to determine the most efficient use of material.*

This feature typically also comes with at least two settings: Soft and Hard.

Soft allocation gives the WorkFace Planning Material Coordinator a chance to juggle the sequencing of FIWPs for the purpose of reviewing different scenarios. Bulk materials soft allocated against a FIWP can be removed and allocated against another FIWP.

Hard allocation removes the bulk material from stock and permanently allocates the materials to a specific FIWP. This is the action that takes place when a FIWP is moved into the four-week look ahead schedule. This ensures that the materials have been allocated against the FIWP and will not be utilized for some other purpose.

For the database to reach this level of data management sophistication there are some important steps upstream from the material receiving function that must take place.

Data integrity:

Quite often the data that leaves the engineering office as "Issued for Construction" must be changed or recreated by the fabricators or vendors.

Steel is a good example of this. The steel model and drawings leave the engineering office without unique identifiers or connection details. The fabricators add the connection details and the piece mark numbers. If the material management database and the WorkFace Planners are going to communicate effectively when talking about steel then they will need to be drawing this information from the same source. At present this is usually a manual extraction from fabricator supplied drawings. Not a very efficient system, we wait until the drawings show up and then build a pick list manually from the drawings and try not to make too many mistakes.

We have seen this information moved electronically from the fabricator to the 3D model and it works quite well, with some preparation. The key piece of preparation is to include electronic data as one of the contractual deliverables between the procurement team and their fabricators. The software used by the detailers produces a CIS/2 file that holds all of the electronic information that we need to populate the 3D model but we normally don't ask for it. So our contracts will need to identify the CIS/2 files as deliverables by CWP and in a sequence that will allow us to populate the 3D model and the material management database before the steel is fabricated. This is possible and has already been executed on several projects.

The same information is usually available for all of the fabricated materials and vendor supplied components. You will also need to standardize the naming convention utilized for the packing slips to ensure that this information is not lost through the shipping process. The description on the packing slip must be an exact replication of the model naming convention. Intelligent 3D models are also being created for vessels and equipment by the vendors but we typically don't ask for them...... So ask.

Following the guide set out on the previous few pages will put you in a position where you know what you need and you know what you have. This will bring you to the next critical fork in the road.

You look at the way that material is amassing in the laydown yard and you see that we have a big pile of cable tray, most of what we need. You also know that you are having trouble getting pipe spools due to some equipment failures in the fabrication facility. Most of the steel is erected so if you wanted to you could get some Electricians and start erecting cable tray. Experience tells you that when you put in cable tray before the pipe is in, you spend a lot of time reworking the cable tray to allow the pipe to fit.

So do you follow the schedule (pipe) or develop workfronts based upon what material you have (cable tray)?

Consider this:

We talked earlier about the Path of Construction and how sometimes you have to step out of sequence due to material shortages. This is quite common and most project managers see this as part of their job description. When progress gets stalled they step in and use their broad view of the project and experience to move on to the next best option. The logic is that the culmination of work fronts being executed will cause the project to reach completion. This is a deviation from the Path of Construction and <u>does not</u> fit well with WorkFace Planning.

WorkFace Planning is about systems, processes and long-term project goals. Our Project Managers are experts in these fields but quite often get drawn into the business of firefighting which allows the fire prevention program to get off track.

It is hard to remember that your goal was to drain the swamp, when you are up to your ass in alligators   :)

Here is the test: You are watching a process and something goes wrong (material does not show up), do you step forward and fix it, or do step backwards to see how the process broke down?

The answer: We have lots of people who can step forward and not many that will step back. WorkFace Planning needs people who can stand back, look at the process and make changes that prevent the next process breakdown.

So the answer to the first question about pipe and cable tray is to find a solution that will improve the supply of pipe, this will bring you long term benefits not short term gains.

Material Management Optimized:

If you have recently ordered anything from the internet and had it shipped through one of the parcel delivery services you probably took for granted the fact that it showed up the next day or the day after. You probably didn't concern yourself with where it came from or how it travelled great distances.

It's a bit like the sun and the moon again. Well why is it that we don't have material management systems on our projects that work the same. It is even simpler for us, we know ahead of time who is going to order what.

Part of the answer is competition, once in place, our material management organizations don't compete with anybody for our business. You don't have a choice in who you order your FIWP from.

In my travels I have spent my fair share of time instructing classes on customer service and the roles and responsibilities of suppliers, customers and clients. The material management systems that we use on construction projects have some fine examples of these roles and show us the difference between the outside world and our internal systems.

Consider this:

In this example think of the Construction Team as the customers and the Material Management Team as the suppliers with the client being the guy who pays the bill.

- The customer needs a service; please supply these materials.
- The suppliers provide the service; Collects materials and then delivers them as requested.
- The client pays the bill.

Three unique groups with distinct needs and expectations.

In the outside world the customer and the client are the same person. You walk into a store, request a service, receive the service and pay the bill. For the supplier, being paid and making profit are directly linked to customer service. In our internal systems the supplier's level of customer service is not related to being paid. The client pays the bill based upon adherence to the contract, not customer satisfaction. This has created a system that allows the material management team to become misaligned with their role. Right now the average material management system can satisfy their contract very well and be paid accordingly, while still not providing the minimum service that WorkFace Planning will require.

So here is the challenge for you contract gurus: Design a contract that drives customer satisfaction. Customers not clients; if construction gets their stuff on time and it allows them to build the project efficiently then the client will be happy.

The other obvious contributor to effective material management is that the folks who run the parcel services are professionals. They move items and manage data for a living and they are very good at it. Typically our material management teams are made up of well meaning, hard working individuals that were trained in some other field, like engineering or one of the trades. This makes them subject matter experts for the components that they are handling but students when it comes to data management. So be aware that a good contract that drives customer satisfaction is still only the starting point. The organizations that we have employed now still have to rise to a new level of data management to satisfy our WorkFace Planning needs.

Ultimately, the optimized material management system will be to hire an organization that does this professionally. Invite them to set up your material management process based upon your inputs and the required outputs and then train your people to manage it.

Another option is to invite the organization to manage materials for you completely. The practice of outsourcing the whole procurement process is becoming more common across the globe as organizations focus on their core competencies. Recently in Australia, a group of large completely diverse companies banded together and formed a procurement team that supplies all of them with their requirements. The logic is that they can become more efficient by outsourcing (internally) with the added bonus that it creates buying power and reduces the overall cost of materials.

So imagine this: Some smart cookie designs a procurement company that scourers the world for suppliers and fabricators, then invites Engineering companies to submit their 3D models for price and schedule quotes. The procurement organization source and deliver the materials while driving the fabricators to maximize efficiencies and reduce costs.

(It's the Wal-Mart plan applied to construction).

If this is one step too far then consider this as an intermediate stage:

Try thinking about the whole process as three different groups that are interdependent. Procurement, Material Management and Warehousing.

The Procurement Team are responsible for the ordering of long lead items, vessels, equipment and bulk materials that will supply the fabrication shops.

The Warehousing Team are responsible for receiving the material at the onsite laydown yard and then the handling of materials until installation.

Material Management is everything in between. This includes the day to day management of the fabrication shops and the module yards, and most importantly: the total management of data.

The total management of data is the glue that holds all of the components together. It should be the common language that allows the bulk order of materials to have association with FIWPs and ultimately the start up of the facility. The base for this common language is the 3 Dimensional model.

The 3D model is the ideal communicator; it takes reams of data, forms it into information (drawings) and then translates that into knowledge (pictures) and ultimately into understanding (colour coding). Concrete understanding from conceptual information.

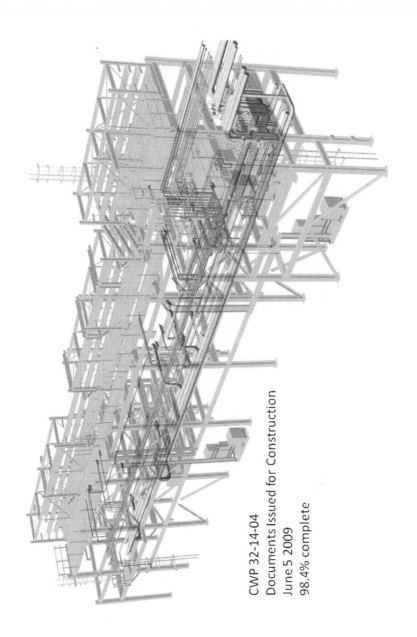

CWP 32-14-04
Documents Issued for Construction
June 5 2009
98.4% complete

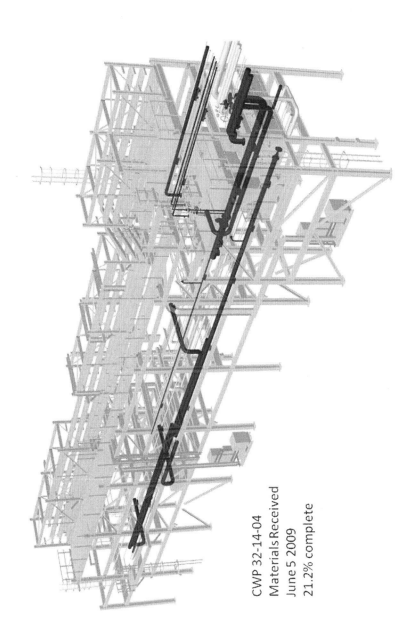

CWP 32-14-04
Materials Received
June 5 2009
21.2% complete

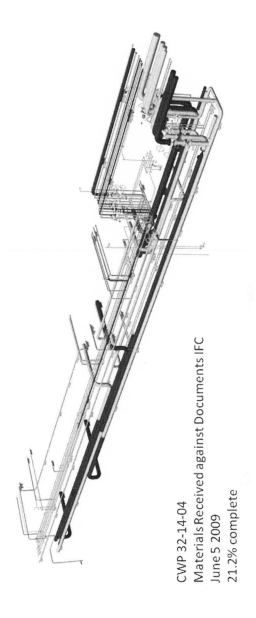

CWP 32-14-04
Materials Received against Documents IFC
June 5 2009
21.2% complete

Imagine the difference in understanding that we convey between these images and a report that just shows that we have 98.4% of the documents and 21.2% of the materials we need for CWP 32-14-04.

The beauty of the 3D model is that it is already the central depository for most of the project data. I think of it as a huge database that gets transformed into pictures.

So why not utilize this project database to manage all of the project's data. It's already there, all we have to do is to develop our software so that we can interact with the data in a manner that makes sense to the user.

And this is what it looks like:

The Document Control Team receives documents against the list in the database and every time that we put a check mark in the received column for a drawing, the drawing in the model changes colour. A counter built into the database tracks the total number of drawings expected (planned) against the total number received (actual) and reports a % complete for each CWP.

Then we take the same process and apply it in the fabrication shop; A material management representative imbedded into the shop, manages the model to show the fabricators what drawings they have. Then dynamically allocates bulk materials against the drawings to show the work that can be started. The operator captures the fabrication completions against the list of drawings and reports progress to the shop and to the project. The fabricators apply barcodes (or RFID tags) to the completed components and the information is entered against the item in the 3D Database. This allows the onsite warehousing team to scan items as they are received and stored. The scanner that that the yard guys use records the GPS location of the item as it is received with the information uploaded into the database across a wireless connection or through a docking port at the end of each day.

Now the project has a very accurate understanding of "*what they have and what they don't have*".

We have applied this process on several projects and enjoyed some great successes, but we never really achieved the level of data integrity that we needed to operate as I just described. (This is reality creeping in again).

There are software programs out there already with great features that can manage information just as I described, but it always comes back to the integrity of data. If you try to apply this process to your existing data, it will look like the software is failing. It takes a lot of design and effort to get any project to the point where the data can function as I described here, but it is well worth it.

I will spend some time on the integrity of data later in the book, for now I want you to know that it can happen and that your level of data management is the key to improved material management performance.

Material Management Summary:

The ineffectiveness of material management is the most common complaint that I hear from projects that achieve poor results. The most common solution applied is for the Construction folks to order everything and then to try to manage it inhouse. This usually compounds the problem of data management and leads to unbelievable situations, like this one:

Towards the end of a recent mega project, we found a situation where the material management group had an excess of raw 8" carbon steel pipe. So they had arranged to sell it back to the supplier at 10% of the original cost (complete with the mill test reports that ensure the quality of the product). At the same time construction were waiting on a shipment (ordered through the material management group) of exactly the same material (from the same supplier) so that they could complete some critical work needed for start up. I am not making this up; those of you who have worked in this environment know that this is both possible and probable. This is a good example of data mismanagement and the sort of thing that can happen when, well meaning, hard working people are hobbled by the lack of an effective data management system.

The industry does have the ability and the tools to change this, we just need permission.

Chapter 9

Information Streams:

In the previous two sections, we looked at the **Basic Principles** of WorkFace Planning, which is really the extended application of the COAA model. Then we looked at **Resources**, which is the next natural stage of development. In this next section, **Information Streams**, we will look at the structure of information that will be necessary to support the Resources and meet the basic needs of the WorkFace Planners.

In a perfect world we would allow the process of WorkFace Planning to develop through each of these stages one project at a time so that when we have three projects behind us we would have a sustainable WorkFace Planning model to continue with.

If you are working with an existing process that must be changed to accommodate WorkFace Planning, this type of gradual change is the best method. If you do have the luxury of building a process from the ground up (starting a new project management company) then you should start with "Information Streams" as your basis for design.

Before we start this section, it is important to note that most of the information that I will address here is already produced on a typical construction project. This section of the book looks at how we can shape and manage that information so

111

that it can be utilized to support the application of WorkFace Planning by the Construction teams.

<u>The structure of information streams that will feed into the onsite resource systems:</u>

The common language established during the *Basic Principles* and *Resource Deliverables* sections of the book was Construction Work Packages (CWPs). This is the common denominator that crosses all boundaries and is meaningful to all of the project stakeholders. So let's start by looking at how a CWP needs to be structured so that it is meaningful to everybody.

The base model for a single CWP:

   ☐ A logical association of work as determined by construction
   ☐ Same geographical boundaries as a Construction Work Area (CWA)
   ☐ Single discipline (Steel)
   ☐ less than 40,000 hours

As a denominator this translates into: **Work – Time – Cost** at a level that is meaningful to all of the project stakeholders:

WORK

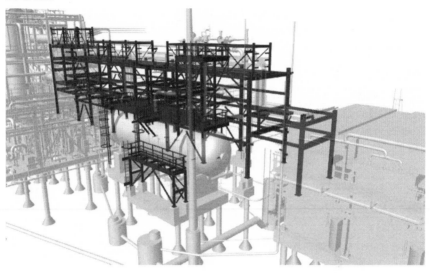

TIME

ID	CWA	CWP	FIWP	Start	Finish	Duration	Jan 2009	Feb 2009	Mar 2009	Apr 2009
1	2955 Pipe Rack									2955 Pipe Rack
2										
3		2955M01 Steel		02/02/2009	27/03/2009	40d		2955M01 Steel		
4										

COST

Cost code: 29-55-M01-12345

(Plant-CWA-SteelCWP-Erection)

I now know that in order to be effective a CWP that supports WorkFace Planning must also work for all of the other project stakeholders and work with all of the existing processes. The following passages are the considerations that our WorkFace Planning project teams used to define the ideal size and shape of CWPs.

<u>Project Stakeholders and components:</u>

The Work Breakdown Structure, Path of Construction, Project Schedule, Engineering, Document Control, Procurement, Fabrication, Material Management, WorkFace Planners,

Field Installation Work Packages, Field level Construction Management, Cost Management, Project Controls & Project Management .

Work Breakdown Structure:

By definition the Work Breakdown Structure is a progressive decomposition of the work to be executed by the project team. The dissection displays the entire work scope through deliverables with parent - child relationships: Plant-CWA-CWP-FIWP.

To satisfy this requirement the CWP needs to be a logical stepping-stone between the CWA (100,000hrs) and the FIWP (1,000hrs).

In recent years the WBS has developed into a tool used primarily for cost management. The alignment of Work, Time and Cost requires that the WBS be the basis for scope definition, the schedule and cost coding.

Path of Construction:

The process of developing the path of construction requires that the project elements be broken down into logical associations of work at a level of definition that can allow the team to sequence the whole project with sticky notes on a white board.

A CWA which has <100,000 hours and boundaries defined by the logical association of work, is the right size and shape for this activity. This model for a CWA also breaks down easily into CWPs for WorkFace Planning.

On a billion dollar project, this might be equal to 10 large pieces of equipment and 10 CWAs. This would give the project team 20 to 30 elements to sequence.

The project team then applies their knowledge and experience to each element and the path starts to develop. The end result from this activity would be the Path of Construction (sequenced) and the basic structure for the level-2 schedule.

The 40,000 hour CWPs are a natural dissection of these CWAs. The CWA has all disciplines, each discipline then becomes a single CWP. Once the CWAs have been sequenced to establish the path of construction the CWPs can then be identified and sequenced, (this forms the basis for the level-3 schedule).

Project Schedule:

The project schedule is the key document on any construction project. It translates the project management expectations into a sequence of tasks that forecast time and eventually cost. In order for the document to be utilized by construction, it also needs to reflect the reality of execution.

When the Construction team is used to establish a standard for the size and shape of CWAs and CWPs, they are adding the reality of execution into the schedule. The schedulers can then concentrate on the mechanics of the schedule based upon the node relationships established by the Construction team.

As described in the last paragraph, the CWAs (developed by construction) become level-2 activities and the CWPs become level-3 activities. The scheduler develops a sequence of activities and a logical association between activities with lag and dependencies. Once established the schedule can then be populated with durations by the Engineering, Procurement and Construction teams.

An important development that WorkFace Planning brings to the schedule is that the schedule should only be developed to a level-3 by the project management team. The WorkFace Planners (Construction) start with this level-3 schedule and then dissect each level-3 activity directly into level-5 activities (FIWPs – a single rotation of work). The scheduler then builds the level-5 schedule from the FIWPs. The rolling wave schedule concept allows construction to develop the level-5 schedule as the level-3 segments (CWPs) are Issued For Construction (IFC).

The existing process for developing schedules requires that the schedulers determine the size and shape of activities and that they apply duration estimates. This only works well if you are lucky enough to have a scheduler who understands all the disciplines of construction (very rare). So this is the weak point in the process, that typically gives us a schedule that looks good on the wall but is meaningless to construction.

The development of single discipline, 40,000 hour CWPs, by construction, allows our schedulers to build meaningful schedules that will be utilized in the field.

Engineering:

Engineering develops based upon the dissection of the whole project into Engineering Work Packages (EWPs). The EWPs need to be small enough to be manageable and need to have a logical association with the schedule, so that the project can be designed in the same sequence as the Path of Construction. Ideally, the EWP is a single discipline that can be designed start to finish by a team of engineers. A single EWP in the 3D model can then be made up of several design areas. This allows several members of the engineering team to work on a single EWP at the same time. The logical step then is to align EWPs with CWPs, so that the design and completion of a single EWP satisfies the needs of a single CWP. This allows each segment of work (CWP), to be designed, procured and constructed in sequence.

$$1 \text{ EWP} = 1 \text{ CWP.}$$

Document Control:

To truly support WorkFace Planning each project must create an environment that is document centric where the Document Control function is responsible for total document management. The ideal system would ensure that the all of the Stakeholders have access to the right drawings at the right time.

To achieve this, the Document Control database should reflect the Work Breakdown Structure and store documents by CWP (the project's common language). The naming convention for all documents would then have to be based upon the WBS. The Project Management Team can then drive this standard back into the departments that generate documents.

For document control teams the world over this is great news, our present systems spit out documents with a wide variety of naming conventions. It is then up to Document Control to translate each document's name so that they can be stored in some logical manner. This leads to a process that captures all of the project's documents but does not allow the end users (the WorkFace Planners) to search for or identify documents by their file name or affiliation.

So the adaption of a project nomenclature that insists upon documents being named by their association with the Work Breakdown Structure will lead to documents that are grouped by CWP. This will allow document control to receive, store and issue documents by their association with a CWP.

Procurement:

The first level of procurement, the identification and acquisition of bulk materials, is not effected directly by WorkFace Planning. The consideration for WorkFace Planning requirements starts with the fabricators and vendors. One of the deliverables from the engineering team is to deliver complete CWPs to the procurement team in line with the CWP release plan. This gives the fabricators and vendors a discernable mandate (supply/fabricate the material required for the CWP) and a delivery date to work towards. The Procurement team then have the responsibility to support and drive these goals.

On traditional projects where the work is not released by CWP the fabricators are free to fabricate based upon their own productivity goals, inches of weld for pipe and tons for steel. This was primarily due to the payment terms established in the contract (paid by inches or tons). The end product of this system satisfied the contract but did not service the needs of construction. A fabricator will set up their shop to fabricate a single size of pipe (8") and do as much as possible before moving on to another size, this is not the way that construction erect. So the project ends up with a yard full of 8" pipe spools while they wait for everything else.

In a traditional environment the procurement team pass documents from engineering to the fabricators without consideration and then try to satisfy their bulk material needs. This activity continues while the construction team are constantly complaining about inappropriate spools being shipped (scheduled for installation 12 months from now). The procurement team sees their role as managing this middle ground, trying to support the production requirements of the fabricators while also giving construction something that they can erect.

The assignment of work by CWP changes all of this to a focus on the needs of the customer (Construction). It should be noted that the process of regulating fabrication is quite common on successful projects. The price we pay is that the fabricators are inconvenienced and that we may need to redefine their contracts. The benefit to the Procurement team is that they can now manage the process and satisfy the needs of construction by issuing work one CWP at a time. In turn this also satisfies the Procurement team's obligation to the level-3 schedule and the Path of Construction.

Fabrication:

As mentioned in the previous passages the process of fabrication can be adversely effected by the execution of work by CWP. This is primarily due to the contracts that we have in place and the payment scale that is meant to drive production rates. As valuable stakeholders in the project, fabricators and vendors must be given consideration. While their need to be productive and make profit are important, it is still secondary to our need to procure materials in a logical sequence that supports construction. So the Path of Construction and the release and completion of work by CWP will need to be identified in Request for Proposals and contracts. So that the fabricators can make allowances for the productivity losses that will be experienced.

Material Management:

The process of managing materials is heavily dependent upon the naming convention used by the fabricators to identify their components. Ideally, this is an exact copy of the naming convention utilized in the 3D model. This allows the WorkFace Planners to extract information from the model and use it to order materials. Therefore, when we have the procurement of materials sequenced by CWP and Construction ordering materials by CWP, it makes good sense to also receive and store materials by CWP. The process of where to store materials becomes further simplified when the warehouse team knows that a group of materials (CWP) will be received over a period of three months and the same materials will be drawn by Construction over a period of two to three months. This could lead to the yard being organized by CWP.

WorkFace Planners:

The ideal scenario for the WorkFace Planners is to receive a single chunk of work (a CWP) at a time and then dissect that piece into FIWPs. That piece of work should be big enough to be a level-3 activity on the schedule and yet small enough to fit into the Planner's head. This makes the ideal size of CWPs 20-40,000 hours. This will break down into 20 - 40 FIWPs, which is a manageable association of elements.

The logic: Below 20,000 hours the work is too detailed for a scheduler to develop and above 40,000 hours it means that the Planners, General Foremen and Superintendents need to manage all of the elements for more than 40 FIWPs in their heads.

Field Installation Work Packages:

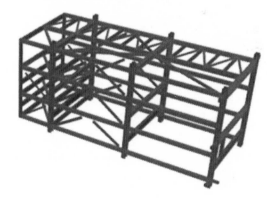

The primary function of the the FIWP is to represent the schedule in small, well-defined segments of work. To do this well each FIWP must be a logical breakdown of a larger component, a CWP. This relationship needs to balance, so that the CWP breaks down easily and a series of FIWPs roll up easily. We can create this balance when we predetermine the size and shape of the components. The ideal balance is to have a 30 to 1 relationship, this creates CWPs that are meaningful to project managers and FIWPs where there is enough definition to clearly identify the constraints.

Field Level Construction Management:

On a typical project the Superintendents, General Foremen and Foremen are responsible to interpret the schedule so that they can coordinate and apply their resources to execute the work. On a project where WorkFace Planning is being applied, this is still true but there is a subtle change in the relationship.

The size and shape of the FIWP is based upon a vision of what the field supervisors believe that a crew will need to be productive. A series of these FIWPs then needs to be a logical association of work that can be constructed in an optimal manner over a period of several months by several crews. The size and shape of a CWP is therefore being determined by the needs of construction and a rolled up collection of FIWPs. A level-3 schedule built with these CWPs then becomes a reflection of the needs of Construction (not the other way around).

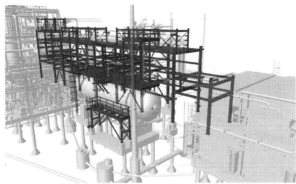

The benefits of this scenario play out when the CWP is first released and the WorkFace Planner pulls these people into a room with the 3D model projected on the wall. The WorkFace Planner extracts the CWP from the rest of the 3D model and then pointing to the projected image he asks, "Which piece should we install first?" Each of the Field Supervisors will then draw on their own experience and start to visualize the sequential assembly of the work. If the CWP is the right size then the field supervisors can start to visualize FIWPs within the boundaries of their resources, (crews and equipment).

Cost Management:

The tracking of cost is the third critical element to understanding the health of any project. A balanced project will have an acceptable level of quality, be aligned with the schedule and be on target to hit the projected cost. In the world of construction, this means that we must understand and track the costs associated with each specific element of work. This is critical for establishing the original estimates and then tracking our performance against the estimate.

For this to be functional we need to have a direct alignment between the work that gets done and the costs that are incurred to do that work. The Work Breakdown Structure is the roadmap for this alignment and is the cornerstone for the dissection of work and the development of the schedule. When the WBS is structured to support the natural dissection of work through Plant - CWA - CWP - FIWP it also serves well as the map for tracking the costs associated with each of these elements.

Sample: Cost code: 29-55-M01-12345

Plant-CWA-SteelCWP-Erection

The Foreman initiates the interaction of the cost codes and the work completed as they apply cost codes to timesheets. This is one of the functions that the FIWP facilitates. Each FIWP is expected to identify a very specific scope of work and then to indicate to the Foreman the cost code that this work should be assigned to. From this point the cost codes can be rolled up to a CWP level or laterally into a commodity (Stainless steel pipe welding) or specific activity (Torqueing bolts).

Typically, a construction project talks about cost as hours mapped against work in the form of a Productivity Factor (Hours Earned/Hours Burned).The time recorded against activities can then be translated into dollars and added to equipment, material and Indirect costs for a Total Installed Cost for each CWP.

This is the point where the alignment of Cost Codes and CWPs pays off for the Project Controls Team. If manhours and equipment costs can be accurately assessed against each CWP then the monthly report to Project Management can show the % of progress recorded against each specific CWP and the % of $ spent against the budgeted amount.

There are several other considerations when tracking costs that are not directly impacted by the size and shape of the CWPs. The Indirect costs of cleaning staff, security, office personnel and management are not directly linked to work that is executed but still need to be tracked and managed against the total direct costs generated from the execution of CWPs.

The rest of the project's costs will receive a residual value from the tracking of direct CWP costs in the form of a baseline that all other costs can be tracked against.

Project Controls:

The project controls team are primarily responsible to summarize the project's performance in the areas of Work, Time and Cost. A standardize CWP that establishes set boundaries for: how much work got done, how long it took and how many hours we spent on it, is the ideal juncture for these elements.

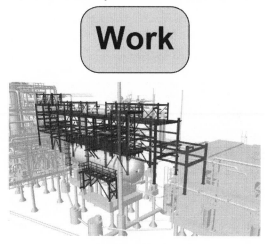

Level -3 activity (approximately 40,000 work hours)

ID	CWA	CWP	FIWP	Start	Finish	Duration	Jan 2009	Feb 2009	Mar 2009	Apr 2009
1	2955 Pipe Rack									2955 Pipe Rack
2										
3		2955M01 Steel		02/02/2009	27/03/2009	40d		2955M01 Steel		
4										

Cost code: 29-55-M01

(Plant-CWA-SteelCWP)

These three elements can then be rolled up and reported against the schedule, the project's key document.

Project Management:

To enable the efficient management of a project the Project Managers must have access to the right information at the right level of definition. (Not too detailed and not too ambiguous).

The correct level of definition is level-3 on the project schedule (CWPs). This allows Project Managers to compare Work, Time and Cost for the entire project and then to zoom in on the elements that are not aligned.

When the size and shape of CWPs is developed to suit WorkFace Planning (<40,000 hours and a single discipline) it also supports this Dashboarding process very well.

Dashboarding is defined as: an information management process that is designed to compile multiple fragments of information into a coherent synopsis that allows users to gauge the project's results against the plan, much like the dashboard in a car.

The key input for this process is standardized information and that is what we get from regulated CWPs and the application of WorkFace Planning.

Note:

For the purpose of Module construction, where we build large modules in a construction environment and then ship them to the site, we have used the same principles laid out here but have altered the definitions to suit the way that modules must be constructed:

A single Module CWP is:

    ▢   A logical association of work, determined by engineering design.
    ▢   Boundaries are set by the maximum shipping envelope.
    ▢   All disciplines

A module number then appears in the WBS as a CWP

FIWPs are then defined by layers and separated by disciplines. All the work on the bottom layer should be completed before the work moves to the next layer. A single crew could work on 4 or 5 different modules in a single rotation.

Information Streams:

The following diagram shows the streams of **Data** that form a river of **Information**. The WorkFace Planner draws from this river of information to develop packages of **Knowledge** that become **Understanding** in the hands of the Foremen.

Data – Information – Knowledge – Understanding.

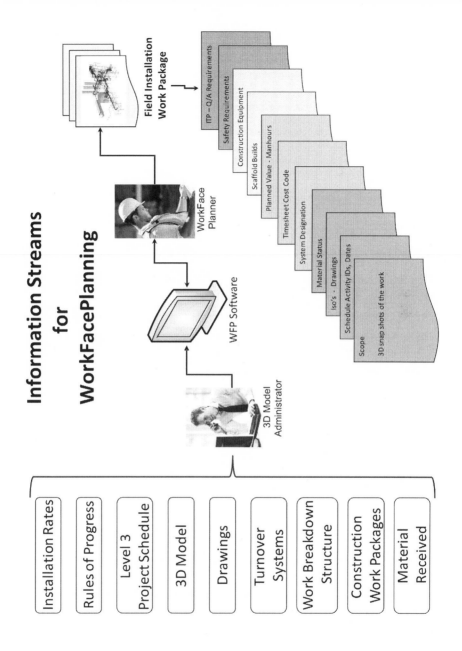

Information Streams for WorkFacePlanning

Field Installation Work Package

WorkFace Planner

WFP Software

3D Model Administrator

ITP – Q/A Requirements
Safety Requirements
Construction Equipment
Scaffold Builds
Planned Value - Manhours
Timesheet Cost Code
System Designation
Material Status
Iso's - Drawings
Schedule Activity IDs, Dates
Scope
3D snap shots of the work

Installation Rates

Rules of Progress

Level 3 Project Schedule

3D Model

Drawings

Turnover Systems

Work Breakdown Structure

Construction Work Packages

Material Received

As described in the previous passage the standard format for information packaging should be a Construction Work Package (CWP). With this established we can now spend some time on the format of the data and the method that we use to move it from its origin to the customer, the 3D model administrator.

Installation rates:

The projected amount of hours it will take to install each cubic meter of concrete, tonne of steel, meter of pipe, meter of cable or individual components: for all commodities. These figures are normally used to develop the estimate and are developed from industry averages or historic performance and then multiplied by site-specific conditions. The tables are a deliverable from Project Management's Project Controls Team to the 3D Model Administrator at the very start of the project. The WorkFace Planning software will draw from these tables when calculating the planned value (PV) of components.

Rules of Progress:

This standard establishes the rate at which the construction team will earn value (EV) from the Installation rates. The construction team may earn 10% of the total Planned Value (PV) for a spool of pipe when it is received on site and then another 30% when it is rigged into place etc. This table is a deliverable from the Project Management's Project Controls Team to the 3D Model Administrator , prior to the start of construction. The WorkFace Planning software will calculate planned value from this matrix based upon the tasks identified by the WorkFace Planner.

Level-3 Project Schedule:

As detailed in earlier passages, the ideal level-3 schedule is made up of Construction Work Packages (CWPs) that each represent a single discipline and less than 40,000 hours. The schedule should also be based on the Path of Construction. The Engineering portion of the schedule should show completion dates for Engineering Work Packages (EWPs) and

1 EWP must equal 1 CWP. The initial level-3 project Schedule is a deliverable from the Project Management's Project Controls Team to the 3D model administrator prior to the start of detailed engineering. The schedule is expected to be updated weekly, for the life of the project, as new information becomes available and progress is added against activities. The Project Controls Team should post the electronic file of the schedule in a designated file folder within the Document Control database each week in preparation for the coming week. The 3D model administrator will draw the current file and apply it to the 3D model through the WorkFace Planning software. This will give the whole project a 4D model based upon the latest weekly schedule.

Rolling wave schedule: The bulk of the schedule would still be presented at level-3, with only the next 8 weeks being represented at a level-5. (The portion that the WorkFace Planners have planned).

Hey how come there is no mention of a level-4 schedule?

I'm glad you asked:

There are several reasons why projects do not need level-4 schedules, the main one is that they are not real. The people who are tasked with the job of dissecting a level-3 into a level-4 are rarely qualified to do it.

Level-3 schedules represent the logical association of work for single disciplines and the EWP/CWP release plan for Engineering. The previous passages talk about the appropriateness of these CWPs from a Project Management perspective. When we elaborate the schedule beyond CWPs we cross the line into the area of construction knowledge and application.

A level -4 activity would need to be about 30% of a CWP. This is the point where we need construction logic to help break out this piece of work so that we maintain a logical association and sequence of the work. To ask our schedulers to do this on paper without a base of field level understanding is where the problem starts.

We could ask our constructors to do this for us, but if they did do it they would still have no use for it. The WorkFace Planners want to receive work in <u>level-3</u> chunks and use it to produce <u>level-5</u> chunks.

The driver for a <u>level-4</u> schedule is usually from Project Management, who would like to develop another level of cost definition. The theory is that it then gives Project Managers and Cost Managers more confidence in the projected costs. The reality is that it complicates the equation and then creates a false expectation. So we spend a lot of time, effort and money on an exaggeration of the truth that we work around for the rest of the project.

The application of WorkFace Planning does not need a <u>level-4</u> schedule, so don't do it and then add the money that you saved to the benefits column.

Back to Information Streams:

3D Model:

The ideal 3D model for WorkFace Planning and the application of WFP software is a direct reflection of The Work Breakdown Structure (WBS). Each CWA is equal to the sum of a number of design areas (to be determined by Engineering). The attributes enabled on each model should be determined by the 3D model administrator from the Project Management Team based upon the needs of WorkFace Planning and the capacity of the WFP software.

As a deliverable: the model should be uploaded to a folder within Document Control each week for the entire duration of the project. The 3D Model Administrators will then draw the new model and update the WFP software prior to the commencement of work each week.

The future of Total Information Management and full-blown WorkFace Planning is going to be 3D model centric. The model will be the great communicator and the entire project will draw from it to answer questions. The path to this standard is for us to maintain the model through construction, turnover and Start up. This means that the engineering team will need to maintain accountability for the model's integrity until Commissioning. The as-built and redline documents will need to be submitted to the onsite engineering model administrators on a weekly basis. The end goal must be to produce a model each week that truly represents the known reality.

In the near term, a simple way to maximize the effectiveness of the model and the information displayed by the WFP software is to ensure that each meeting room has a live model projected onto the wall for <u>every single meeting</u>. Joysticks that allow the users to cruise around the "read only" model and laser pointers will then make your meetings come to life. I have seen this applied and it will save you thousands of hours both in the time spent in meetings and the clarity of understanding that is produced.

Documents:

The Engineering team are responsible to develop and maintain the Document Control function for the entire project. The Document Control function must be a single department that receives documents from Engineering and provides documents to the Construction Contractor through the 3D model Administrator and the WFP software. The ideal situation is to hyperlink objects in the 3D model to their root document in the Document Control database. This allows the WorkFace Planners to "right click" an object in the 3D model and view/print the latest revision of the drawing in the Document Control database from their own desktop.

To facilitate this the Document Control database must be online and must be constructed based upon the Work Breakdown Structure (WBS).

As the hard copy drawings are submitted to Document Control from Engineering, there must also be an electronic copy (IDF/PCF) from the 3D model. This allows the 3D Model Administrator to load the WFP software and display all of the "Issued for Construction" (IFC) drawings in the 3D model.

Truly functional projects are all document centric. Please do not consider this one of the functions that is optional. In all of the elements that we have talked about there is not a good one that you can leave out of your model for WorkFace Planning, but this is definitely a bad one to not do.

Turnover Systems:

Prior to the start of Construction the Ready For Operations Team (RFO) are responsible to identify exactly how they would like to receive the completed plant. They do this by identifying the turnover systems on Piping and Instrumentation diagrams (P&IDs) and building a sequence of turnover between the systems. The engineering teams are then responsible to identify the systems in the model with the end result being that every component is identified primarily by it association with the Work Breakdown Structure (WBS) and then by association with a turnover system.

The 3D Model Administrator receives the system allocations through the 3D model and links the system components through the WFP software. This allows the WorkFace Planners to view turnover systems in preparation for testing and system completions. The WorkFace Planners can then support the Turnover Coordinator by utilizing the WFP software to display systems and progress.

The Work Breakdown Structure:

The Work Breakdown Structure (WBS): A progressive dissection of project deliverables with parent – child relationships.

The WBS is typically established prior to detailed engineering and is the base platform for the 3D model, Document Control, Material Management, the Project Schedule, Project Controls and Cost coding. This allows the WorkFace Planners to align work with time and cost.

The size and shape of the WBS elements is very important to the application of WorkFace Planning. Typically the dissection should look like this:

Plant/Train

- Construction Work Area (CWA)
- Construction Work Package (CWP)
- Field Installation Work Package (FIWP)
- Drawing
- Component

The ideal shape and size of these elements is:

CWA: Geographic cube of logically associated work – all disciplines – less than 100,000 hours = to a single level 2 schedule activity.

CWP: Same boundaries as the CWA – single discipline – less then 40,000 hours = to a single level-3 schedule activity.

FIWP: one rotation of work for one crew – single discipline – less than 1000 hours = to a single level-5 schedule activity.

Developing a standard for the WBS and then insisting that it be utilized across every department of your project will give you a common language that will enable communication. The payoff comes when the WorkFace Planner identifies a component for installation. The common language established will allow Material Management, Document Control, Project Controls, Engineering, Project Management and Construction to have a common understanding of what is about to happen.

Construction Work Packages:

As covered in the initial portion of this chapter, the Construction Work Packages (CWPs) are a deliverable from Engineering to Document Control that are identified on the level-3 project schedule. A CWP can only be considered complete when more than 95% of the drawings have been issued for construction and submitted to Document Control.

The engineering portion of the CWP (part A) is then flagged as complete by Document Control to the WorkFace Planners. The 3D model Administrator applies the electronic files to the 3D model through the WFP software and validates that the drawings match the parameters of the CWP in the 3D model.

Part B of the CWP is the confirmation that the material requirements have been met. A weekly report from the Material Management team will show the WorkFace Planners the received status of every CWP. This function is one of the deliverables from the Material Management Database.

Part C of the CWP is developed by the CWP Coordinator for the Construction Team and consists of the QC and Safety requirements along with a roll up of the resources identified by the WorkFace Planners (scaffold, equipment and craft).

And once again: The ideal shape and size of a CWP is a single discipline with less than 40,000 hours. The WorkFace Planners then take this area of work and utilize the WFP software to dissect the work into Field Installation Work Packages (FIWPs), with guidance from the General Foreman and Superintendent. A single FIWP will then represent one rotation of work for one crew, contain approximately 1000 hours and become a single level-5 schedule activity.

In order to facilitate this process a single CWP must equal a single Engineering Work Package (EWP) and appear as a single level-3 schedule activity. The delivery sequence of the CWPs must be based upon the Path of Construction. Materials must be procured and delivered to site by CWP and appear in the project schedule as such.

The initial delivery of the CWP from Engineering is through the 3D model to the 3D model Administrator. Each CWP is then monitored by the CWP Coordinator, who utilizes the WorkFace Planning software to develop weekly status reports on all of the project's CWPs.

Material Management:

The Material Management Database should deposit a material received file to a designated folder in the Document Control Database each week. The 3D Model Administrator picks up the file as part of the weekly WFP software update. This allows the WorkFace Planners to overlay CWPs and FIWPs in the 3D model with the material received information and produce a material status report for each CWP and FIWP. As illustrated in the Material Management Flow Chart. The status of material is the primary constraint for all FIWPs and is the trigger for the assembly of the Hard Copy FIWP.

For this process to be functional, the Material Management Database must be structured to receive individual components by their 3D model naming convention. This must be derived from the Work Breakdown Structure.

To ensure that this is accurate the Material Management Database should be populated from the 3D model by engineering each week for the life of the project.

Chapter 10

For the first 8 chapters of the book we were concentrating on the application of WorkFace Planning, then with Chapter 9 we looked at where the information must come from in "Information Streams". The following chapter is the next level of thinking and application. If you are committed to achieving world-class results through the application of WorkFace Planning then you will need to build a step-by-step action plan that looks something like this:

WorkFace Planning by Design

If you have made it this far in the book then you probably have a head full of ideas and concepts that seemed to make sense when you read them, but there is probably no natural flow or web of connections. At least that's what happens to me, whenever I start talking about what could be. So in order to try to bring this whole thing together and give you some chance of one day explaining it to others, I have developed a master flow chart – WorkFace Planning by Design. This is the step-by-step guide on how to implement WorkFace Planning –nose to tail.

I designed and constructed the following flow chart with the expectation that it will probably never be applied the way that it is displayed. I want it to be a starting place for you, a generic sample that you can adapt to your application of WorkFace Planning. My logic and experience tells me that the closer that you can stay to this model, the better your results will be.

The ideal outcome from this chapter of the book would be for you to have a fairly good idea of the effort that it will take to make this work. And then to initiate the process by printing this flow chart (from the web site) and posting it onto the wall. Then right under it build another flow chart that shows how your organization works now. Look for consistencies between the two and identify where the easy wins might be.

The final element is to understand how much change your organization can handle and how much change (pain) you want to impose, then get started.

This is also a good time to remember Albert Einstein's rule for process development:

"We can't solve problems by using the same kind of thinking we used when we created them."

So don't look at WorkFace Planning from where you are today,

Visualize your full-blown application of WorkFace Planning, several years from now and then look back at your organization.

Two things that I know for sure about WorkFace Planning:

It is not quick and it is not easy.

The WorkFace Planning by Design flow chart is linked to the node descriptors on the pages that follow directly after it. If you have access to the internet, you may want to read this next section from the website. There the nodes on the flow chart expand to the descriptions of the activities and then to the third level of detail and other examples.

Stakeholder swim lanes: For the purpose of this book, I have developed these descriptions of the following stakeholders.

 Owner

Owner: The Client, an organization that initiates a project by developing a business need, the eventual operator and owner of the process unit that is the deliverable of the project.

The Owner is not the Project Management team but may appoint personnel to the Project Management team to facilitate an understanding of operational requirements.

Project Management

Project Management Team: The organization that is responsible to manage the project from concept to Ready for Operations. This group is usually a third party organization that specializes in Project Management. It may be an EPC company or a division of the Owner's organization that is specifically dedicated to managing projects for the owner. The Project Management Team has total accountability for the management and success of the project. They have representatives from the Construction Management Team, Engineering, Procurement, Safety, Quality WorkFace Planning and Operations.

Engineering

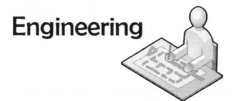

A third party organization that reports directly to the Project Management team. They are responsible to elaborate the owner's concepts into a detailed design. This organization is directly responsible for the functionality of the process and the generation of project information.

Procurement

A third party organization that can be specialists in procurement or an arm of the Engineering organization. This team is responsible to engage fabrication contractors and procure bulk materials.

Material  Management

May be a third party organization that specializes in the management of material, an extension of the Procurement organization or an arm of the Project Management team. They are responsible for the management of fabricators, pipe, steel and electrical, but are not responsible for the fabrication of vessels and equipment. The organization is responsible for the onsite warehousing management and distribution of all materials.

Construction

Construction Contractor: A third party organization engaged to construct the project. They are sometimes referred to as the General Contractor. Responsible to engage and manage sub contractors as required.

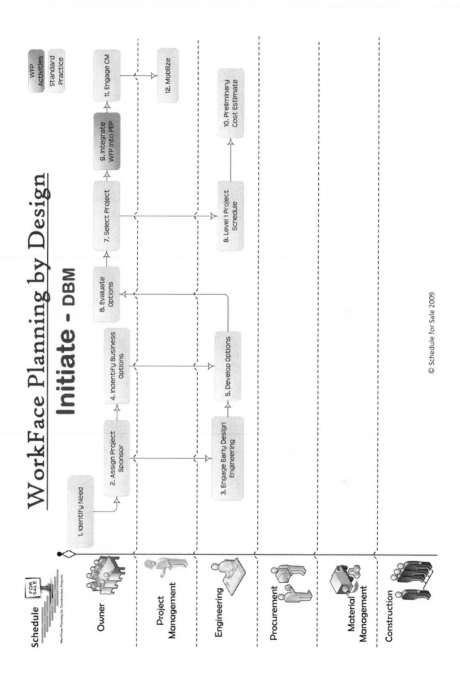

WorkFace Planning by Design

Initiate - DBM

- Schedule — WorkFace Planning for Construction Projects — FOR SALE
- Owner
- Project Management
- Engineering
- Procurement
- Material Management
- Construction

1. Identify Need
2. Assign Project Sponsor
3. Engage Early Design Engineering
4. Indentify Business Options
5. Develop Options
6. Evaluate Options
7. Select Project
8. Level 1 Project Schedule
9. Integrate WFP into PEP
10. Preliminary Cost Estimate
11. Engage CM
12. Mobilize

WFP Activities
Standard Practice

© Schedule for Sale 2009

WorkFace Planning by Design

Plan 1 – EDS - FEED

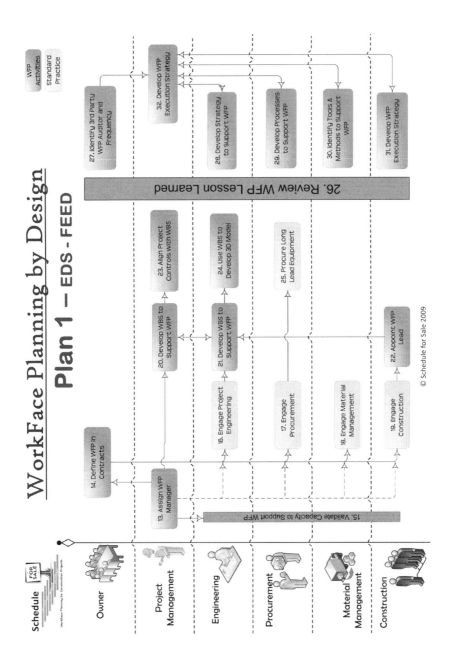

© Schedule for Sale 2009

WorkFace Planning by Design

Plan 2 – EDS - FEED

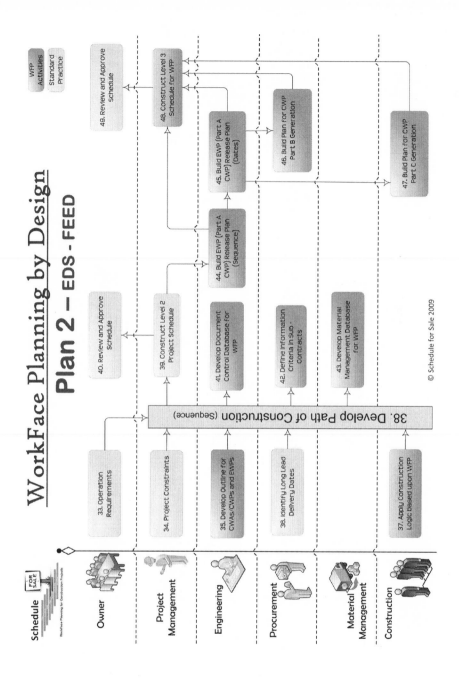

33. Operation Requirements

34. Project Constraints

35. Develop Outline for CWAs/CWPs and EWPs

36. Identify Long Lead Delivery Dates

37. Apply Construction Logic based upon WFP

38. Develop Path of Construction (Sequence)

39. Construct Level 2 Project Schedule

40. Review and Approve Schedule

41. Develop Document Control Database for WFP

42. Define Information Criteria in Sub-Contracts

43. Develop Material Management Database for WFP

44. Build EWP (Part A CWP) Release Plan (Sequence)

45. Build EWP (Part A CWP) Release Plan (Dates)

46. Build Plan for CWP Part B Generation

47. Build Plan for CWP Part C Generation

48. Construct Level 3 Schedule for WFP

49. Review and Approve Schedule

WFP Activities

Standard Practice

Schedule

Owner

Project Management

Engineering

Procurement

Material Management

Construction

© Schedule for Sale 2009

WorkFace Planning by Design

Execute – Detailed Engineering

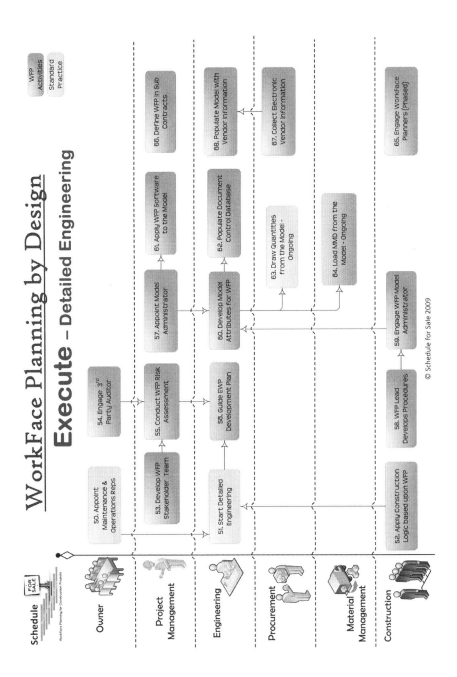

WorkFace Planning for Construction Projects

Schedule

Owner

Project Management

Engineering

Procurement

Material Management

Construction

WFP Activities

Standard Practice

50. Appoint Maintenance & Operations Reps

53. Develop WFP Stakeholder Team

51. Start Detailed Engineering

52. Apply Construction Logic based upon WFP

54. Engage 3rd Party Auditor

55. Conduct WFP Risk Assessment

56. Guide EWP Development Plan

58. WFP Lead Develops Procedures

57. Appoint Model Administrator

60. Develop Model Attributes for WFP

59. Engage WFP Model Administrator

61. Apply WFP Software to the Model

62. Populate Document Control Database

63. Draw Quantities from the Model - Ongoing

64. Load MMD from the Model - Ongoing

66. Define WFP in Sub Contracts

68. Populate Model with Vendor Information

67. Collect Electronic Vendor Information

65. Engage WorkFace Planners (Phased)

© Schedule for Sale 2009

WorkFace Planning by Design

Execute – Construction

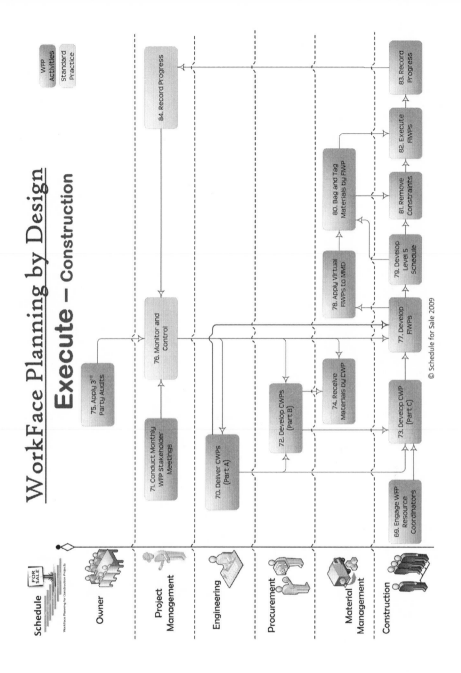

© Schedule for Sale 2009

Legend:
- WFP Activities
- Standard Practice

Swimlanes:
- Schedule — Workface Planning for Construction Projects — FOR SALE
- Owner
- Project Management
- Engineering
- Procurement
- Material Management
- Construction

Activities:
- 69. Engage WFP Resource Coordinators
- 70. Deliver CWPs (Part A)
- 71. Conduct Monthly WFP Stakeholder Meetings
- 72. Develop CWPs (Part B)
- 73. Develop CWP (Part C)
- 74. Receive Materials by CWP
- 75. Apply 3rd Party Audits
- 76. Monitor and Control
- 77. Develop FIWPs
- 78. Apply Virtual FIWPs to MMD
- 79. Develop Level 5 Schedule
- 80. Bag and Tag Materials by FIWP
- 81. Remove Constraints
- 82. Execute FIWPs
- 83. Record Progress
- 84. Record Progress

WorkFace Planning by Design

Close - Turnover

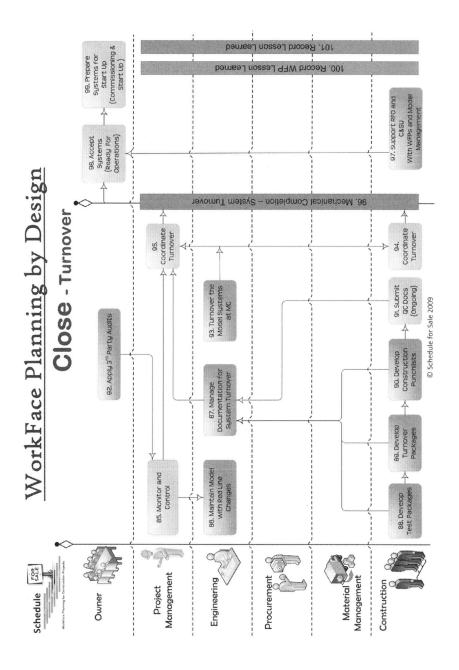

© Schedule for Sale 2009

Initiate: DBM (Design Based Memorandum)

1. **Identify Need:** Owner: The Owner identifies a business need.
2. **Assign Project Sponsor:** Owner: Executive Project Sponsor is assigned to develop the business need into a project.
3. **Engage Early Design Engineering:** Engineering: Engineering Company (usually in house) are engaged to develop a preliminary design for the project.
4. **Identify Business Options:** Owner: Project Sponsor will develop several options to satisfy the identified business need.
5. **Develop Options:** Engineering: Early Design Engineering Team develop the options identified by the Project Sponsor.
6. **Evaluate Options:** Owner: Project Sponsor evaluates the options developed by the Early Design Engineering Team.
7. **Select Project:** Owner: Project Sponsor selects one option from the options developed by the Early Design Engineering Team.
8. **Level-1 Project Schedule:** Engineering: Early Design Engineering Team develop the chosen option to a level-1 definition.
9. **Integrate WFP into PEP:** Owner: If the project is of sufficient cost and complexity the Project Sponsor includes the application of WorkFace Planning as a core component of the Project Execution Plan. The Owner should utilize a WorkFace Planning Consultant or in house construction expertise to answer this question. See WorkFace Planning Suitability
10. **Preliminary Cost Estimate:** Engineering: Early Design Engineering Team develop the initial level-1 cost estimate.
11. **Engage CM:** Owner: Owner selects a Project Management Organization to manage the project. This could be an external organization or an internal Project Management Team.

12. **Mobilize:** Project Management Team: Owner invites the PMT to establish the project office and engage the key personnel.

Plan 1: EDS, FEED (Early Design Stage, Front End Engineering Design)

13. **Assign WFP Manager:** Project Management: The WorkFace Planning Manager is engaged and mobilized to the project office. See: <u>WFP Manager Job Description</u>.
14. **Define WFP in Contracts:** Owner: Prior to the initial Request For Proposal, the owner must establish the requirement for WorkFace Planning in each of the major contracts: See <u>Contract Language.</u>
15. **Validate Capacity to Apply WFP:** Project Management: The WorkFace Planning Manager will work with the Owner and the Project Management Organization to evaluate the capacity of each of the organizations being considered for the project. The evaluation will be scored with each organization being rated on a scale of 1 – 100. The final score will be considered a major component in the technical evaluation for each organization: See Evaluation Criteria.
16. **Engage Project Engineering:** Engineering: Successful applicant is offered a contract and mobilized.
17. **Engage Procurement:** Procurement: Successful applicant is offered a contract and mobilized.
18. **Engage Material Management:** Material Management: Successful applicant is offered a contract and mobilized. This could be the same organization as the procurement team or a separate organization that specializes in the management of material.
19. **Engage Construction:** Construction Contractor: Successful applicant is offered a contract and mobilized. The engagement of the constructor is not common at this stage of the project, however this is a key to developing a construction driven project.

20. **Develop WBS to Support WFP:** Project Management: The WFP Manager along with the project Controls team (PMT) will work with the Engineering Team and the WorkFace Planning Manager (Construction) to develop the size and shape of the components in Work Breakdown Structure based upon the PEP.
    See Work Breakdown Structure: Components.
21. **Develop WBS to Support WFP:** Engineering: The Engineering Team will work with the WFP Manager, Project Controls(PMT) and the WFP Lead (Construction) to develop the Work Breakdown Structure based upon the criteria set out in the Project Execution Plan
    See Work Breakdown Structure: Components
22. **Appoint WFP Lead:** Construction Contractor: The WorkFace Planning Lead for the Construction Contractor will be engaged and then participate in the process of defining the WBS.
    See WFP Lead, Construction Contractor Job Description.
23. **Align Project Controls with WBS:** Project Management: The Project Controls Team within the Project Management Organization will develop the schedule and cost definitions based upon the WBS. This will lead to the alignment of Work, Time and Cost. This activity will precede the development of quantities and work-hour estimates.
24. **Use WBS to develop 3D model:** Engineering: The Engineering Team establishes the rules for the 3D model development based upon the WBS. This ensures that the level-3 components of the WBS can be isolated in the model for the alignment between Work, Time and Cost.
25. **Procure long lead items:** Procurement: Procurement Team work with the Engineering Team to secure a basic design on long lead equipment and then place orders with the fabricators.
26. **Review WFP Lesson Learned:** Project Management: The WorkFace Planning Manager (PMT) will lead this one-day session where all of the Project Stakeholders are introduced to the functional application of WorkFace Planning. The WFP Manager will introduce Lessons

Learned from other projects and point out where the issues developed and what can be done to learn from the experiences. The deliverable from this session will be a common understanding of the process and a template for the WorkFace Planning Execution Strategy.

27. **Identify Third Party Auditor and Frequency:** Owner: The owner representative will identify a third party WorkFace Planning auditor and then develop a plan for the frequency of the audits.

28. **Develop Strategy to Support WFP:** Engineering: Working with the WFP expectations from the contract and the information from the lessons learned session, the engineering team will develop a strategy to support WorkFace Planning. This document will include the broad application of the WBS, the size and shape of CWAs, CWPs, and EWPs and the plan to build and execute the CWP release plan. The engineering group will also develop a separate strategy to detail the development and application of the Document Control process and Database that will satisfy their contractual requirements for WFP support. See: Document Control model

29. **Develop processes to support WFP:** Procurement: The Procurement Team will develop a set of processes that satisfy the contractual requirements to support WorkFace Planning with electronic vendor information, adherence to the WBS naming convention and support for the CWP release plan.

30. **Identify Tools and Methods to Support WFP:** Material Management: The Material Management Team will indentify the database and the functionality that will be used to satisfy their contractual obligations for WorkFace Planning Support and address any issues identified during the lesson learned session. The strategy will include the alignment between the Material Management Database and the WBS.
See: Material Management model

31. **Develop WFP Execution Strategy:** Construction Contractor: Based upon the lesson learned session, the processes presented in the bid proposal and the

contractual obligations the WorkFace Planning Lead for the Construction Contractor will develop a strategy for the development of the WorkFace Planning Execution plan. The plan will show the application of WFP software with a list of deliverables from Engineering, Procurement, Document Control, Project Controls and Material Management. The plan will also include the processes for developing FIWPs with details around the execution, monitoring, control and reporting of work.
See: WorkFace Planning Application- Constructor

32. **Develop WFP Execution Strategy:** Project Management: The WFP Manager will develop a WFP Execution Strategy based upon the input from Activities 27-31. (The Stakeholder Strategies). The WFP Execution Strategy will be reviewed and endorsed by the Project Management Team and then distributed to all of the project stakeholders. This document will be the guiding document for the development of WFP execution plans and the application and support of WorkFace Planning for the life of the project. Changes to the requirements laid out in the document can only be made by the Project Management Team.
See: Sample WorkFace Planning Strategy

Plan 2: EDS, FEED (Early Design Stage, Front End Engineering Design)

33. **Operation Requirements:** Owner: Owner representatives establish the basic requirements for the turnover of systems and trains.

34. **Project Constraints:** Project Management: PMT develop a list of project constraints. This would include access to site, environmental restrictions, shipping windows and all other major influences on the project.

35. **Develop Outline for CWAs/CWPs and EWPs:** Engineering: The Engineering team will develop the generic definition for CWAs, CWPs and EWPs based upon the WBS.

36. **Identify Long Lead Delivery Dates:** Procurement: Procurement team provides delivery estimates for long lead equipment and bulk materials.

37. **Apply Construction Logic based upon WFP:** Construction Contractor: Currently applied on some projects as a constructability review. This activity is a cornerstone requirement for WorkFace Planning. The logic of field constructability based upon WorkFace Planning must be one of the key drivers for establishing the Path of Construction.

38. **Develop Path of Construction:** Project Management: Logical sequence of construction is established. As an element of this WorkFace Planning work flow, this process will remain the same except that the size and shape of the components will be mandated and the process will be influenced by the Construction Contractor. The deliverable from this activity will be a sequence of level 2 activities that cover the entire scope.

39. **Construct level 2 project Schedule:** Project Management: Project Controls Team for the PMT develop a level 2 schedule. The basis for the schedule is the sequence developed during the Path of Construction. The Project Controls team then utilizes the Project Stakeholders to apply durations to the activities.

40. **Review and Approve Schedule:** Owner: Project Management Team submits the level 2 Schedule to the Owner for review and approval.

41. **Develop Document Control Database for WFP:** Engineering: The Engineering Team will utilize the WFP Execution Strategy to develop and execute a plan to a build a Document Control Database. The deliverable from this activity will be an online database that is based upon the WBS, which is accessible to all of the project stakeholders. Access privileges for read only, print, search and submit will be designed to facilitate all of the Stakeholders.
See: Document Control model

42. **Define information criteria in sub contracts:** Procurement: Utilizing the standards established in the WFP Execution Strategy, the Procurement Team will develop contract

language that identifies the electronic information, the delivery method and schedule for each vendor.
See: Vendor information minima

43. **Develop Material Management Database for WFP:** Material Management: The Material Management Organization will develop the Material Management Database based upon the criteria set out in the WFP Execution Strategy. The structure of the MMD will be based upon the WBS, have the ability to be populated from the 3D model and have dynamic allocation. The Database must be online and be available to all of the project stakeholders for user defined searches and reports. Access privileges will be designed to facilitate all of the Stakeholders.
See: Material Management model

44. **Build EWP (Part A) Release Plan (Sequence):** Engineering: The Engineering Team will utilize the level 2 schedule to develop a sequence for the CWP part A (the engineering portion) release plan. The CWP release plan will then be used to develop the EWP development sequence, based on the logic that 1EWP = 1CWP.

45. **Build EWP (Part A) Release Plan (Dates):** Engineering: The Engineering Team will apply duration estimates to the EWP development plan and then apply the completion dates to the CWP release plan. This will form the basis for the level-3 schedule.

46. **Build Plan for CWP (Part B) Generation:** Procurement: The Procurement Team will take the CWP release plan developed by Engineering during Activity # 45 and add duration estimates for the procurement and fabrication of materials to each CWP. This information will be identified as a level-3 activity on the project schedule The Procurement Team will then develop a plan and process for adding the procurement portion of the CWP (part B) to each CWP.

47. **Build plan for CWP (Part C) Generation:** Construction Contractor: The CWP coordinator for the Construction Contractor will take the CWP release plan from Engineering and add the estimate for the development and addition of part C (Quality and Safety) to each CWP. This information will be identified as a level-3

activity on the project schedule running in parallel to the development of Part B of the CWP.

48. **Construct** level-3 **Schedule for WFP:** Project Management: This is an activity common to most projects. The application of WorkFace Planning adds these activities to the process: The Project Controls Team within the PMT will progressively elaborate the level 2 schedule to become the level-3 schedule based upon the CWP release plan from Engineering. The submissions for Parts B and C development from Procurement and Construction will also be added to the level-3 schedule as parallel activities. The duration estimates for construction will be developed based upon the project schedule with consideration for project constraints and estimated quantities.

49. **Review and Approve Schedule:** Owner: Project Management Team submits the level-3 Schedule to the Owner for review and approval.

Execute: Detailed Engineering

50. **Appoint Maintenance & Operations Reps:** Owner: Owner will assign representatives to develop systems for turnover and influence the design based upon operation experience.

51. **Start Detailed Engineering:** Engineering: Engineering team is mobilized to start detailed engineering.

52. **Apply Construction Logic based upon WFP:** Construction Contractor: This process is currently applied on some projects as constructability. The application of WFP broadens this role to include having a guiding influence on the size and shape of EWPs.

53. **Develop WFP Stakeholder Team:** Project Management: The WFP Manager (PMT) and the WFP Lead (Construction Contractor) will develop a WFP Stakeholder team with representatives from all of the major stakeholders (each swim lane) and Document Control, Project Controls (PMT and Construction), QA/QC, Safety, Turnover and Field level Construction (Foreman). The team will

meet each month and address issues that impede the application of WFP.
See: <u>Stakeholder Team</u>

54. **Engage Third Party Auditor:** Owner: The owner will engage the third party auditor identified in activity #27 and invite them to join the Stakeholder team for the WFP Risk review as a participant.

55. **Conduct WFP Risk Assessment:** Project Management: The WFP Manager will convene a one-day WFP risk review as the second meeting of the WFP Stakeholder team. The WFP Manager will facilitate the meeting with the Stakeholders and the third party Auditor. The group will identify risks to the application of WFP, develop a probability matrix, build an impact assessment and then list mitigation strategies. The final assessment will be submitted to the PMT for approval.

56. **Guide EWP Development Plan:** Engineering: Engineering Team develop EWPs in line with the <u>level-3</u> schedule. The difference for the application of WFP is that strict adherence to the path of construction will be dominant over production quantities. In addition, there is a strong possibility that there will be mitigation strategies from the WFP risk review that will influence the usual Engineering processes. Progress against the CWP release plan will be monitored and controlled by the PMT.

57. **Appoint Model Administrator:** Project Management: The WFP Manager will engage a 3D model manager as a subordinate.
See: PMT Model Administrator - Job description.

58. **WFP Lead Develops Procedures:** Construction Contractor: The WFP Lead will develop a set of project procedures that reflect the standards established in the WFP Execution Strategy (activity #31) and the WFP Execution Strategy (activity # 32) with the added detail of flow charts and organizational structure.

59. **Engage WFP Model Administrator:** Construction Contractor: The WFP Lead will engage a 3D model administrator as a subordinate. This person will be assigned to work for the PMT Model Administrator in

the engineering house during the engineering stage and then with the Construction Contractor's WFP department as the project moves to the field. See Construction Contractor Model Administrator – Job description

60. **Develop Model Attributes for WFP:** Engineering: The Model Administrator for the Engineering team will work with the Model Administrators from the PMT and the Construction Contractor to develop a set of standard attributes based upon the needs of WFP.

61. **Apply WFP software to the Model:** Project Management: After consulting with the Model administrators, the WFP Manager will select a WFP software that is suitable for the project, with consideration for each of the Stakeholder's needs. The Model Administrators (PMT, Engineering and Construction Contractor) will apply the software to the model and develop a list of deliverables that will be necessary to populate the software. The WFP Manager and the WFP Stakeholder Team will be responsible to develop processes that will ensure the delivery of this information.

62. **Populate Document Control Database:** Engineering: The Engineering Team and the Document Control manager will develop a process that populates the Document Control database with the estimated number of drawings for each CWP. This weekly process would draw estimated drawing quantities from the model and update the CWP estimates in the Document Control Database. The estimate for total quantities by CWP will be the basis for earning progress.

63. **Draw Quantities from Model – Ongoing:** Procurement: Model administrator for the Engineering team draws total quantities of bulk materials from the model. The Procurement Team uses these figures to calculate estimates and order the bulk material from the suppliers.

64. **Load MMD from Model - Ongoing:** Material Management: The Material Management Database is populated at several intervals with IFC'd components

from the 3D model. For the purpose of WFP support this is a critical item. The naming convention established in the model and the WBS must be duplicated exactly in the MMD. At the conclusion of the population activity, the Required At Site (RAS) dates necessary to support the scheduled execution of the CWPs should be added to the MMD. The RAS date is the date when 100% of the material should be received into the onsite Material Management yard, (not the date when it needs to be in the hands of the constructors – ROS date)

65. **Engage WorkFace Planners (Phased):** Construction Contractor: The WFP Lead will engage and train WFPs by discipline in the latter stages of Detailed Engineering and the early stages of fabrication. The WFPs should be initially assigned to the engineering house and then to the fabrication facility before moving to site 8 weeks prior to the start of their discipline's construction as per the WFP Execution Strategy.

66. **Define WFP in Sub-contracts:** Project Management: The Contract team within the PMT must define the need to support or apply WFP in all of the subcontracts prior to Request For Proposal (RFP). The WFP Manager should be used as a resource to help develop the contract language based on the scope of the work. See Contract language.

67. **Collect Electronic Vendor Information:** Procurement: The Procurement team with support from the Construction Contractor's WFPs and model administrators are responsible to collect the electronic information detailed in the procurement contracts during activity #42. This information should be collect weekly and delivered to the Engineering model administrator for incorporation into the 3D model.

68. **Populate Model with Vendor Information:** Engineering: The 3D Model Administrator receives fabrication and vendor information each week and then uploads that information into the 3D model. This may include some translation. If the minimum requirements established during activity #60 have not been met then the

model admin will request the additional information directly from the fabricator/vendor with notification to the contract administrator and the procurement manager.

Execute: Construction

69. **Engage WFP Resource Coordinators:** Construction Contractor: The WFP Lead will engage resource coordinators for scaffold, construction equipment and materials as per the WFP Execution Strategy. This will initiate the development of processes and data management systems for resource management : See Resources – Schedule for Sale

70. **Deliver CWPs (Part A):** Engineering: The Document Control Manager will announce the delivery of each CWP (part A) when the documents submitted as IFC exceed 95% of the estimate drawn from the model. This will be in the form of an e-mail that notifies all of the project stakeholders that the documents are now available through their online connection to the database. This action will constitute delivery of the CWP. It is then up to the stakeholders to print the CWP from the document control database as they see fit. The document control database will date stamp each document with the day that it was printed.

71. **Conduct Monthly WFP Stakeholder Meetings:** Project Management: The WFP Manager will continue to facilitate the WFP Stakeholder meetings each month for the life of the project.

72. **Develop CWPs (Part B):** Procurement: Upon notification of the CWP (part A) completion, the Procurement team will print and deliver the IFC documents to the Fabricator/Vendor. The Procurement team is expected to manage the fabrication of materials by CWP to support of the Path of Construction. The Procurement team will develop CWP (part B) in parallel to the Fabrication/Vendor activities as per their plan initiated during activity # 46 and in line with the level-3 schedule. The documents and electronic information generated by the fabricators/ vendors will be submitted

to Document Control as it becomes available and be stored as an element of the CWP. The documents that make up the complete CWP Part B must be delivered from Procurement to Document Control to satisfy the schedule. See Procurement Flow chart.

73. **Develop CWPs (Part C):** Construction Contractor: The CWP coordinator for the Construction Contractor will develop Part C for the CWP utilizing their QC, Safety and Discipline Representatives. The combination of part A (Engineering), part B (Procurement) and part C (Construction) will constitute the entire CWP.

74. **Receive Materials by CWP:** Material Management: The onsite Material Management Team will receive materials against the line items in the MMD (populated from the model). The material received will be tracked against the total estimated quantities for each CWP. This will produce a % received for each CWP. Dynamic allocation will be prioritized by the sequence in the level-3 schedule.

75. **Apply Third Party Audits:** Owner: The owner will invite the third party auditor to conduct an audit in the first 8 weeks of Construction mobilization and then again at 30% construction complete as per the WFP Execution Strategy.

76. **Monitor and Control:** Project Management: PMT collect progress information from all of the Stakeholders and manages the project to the Path of Construction and the level-3 Schedule. The WFP fundamentals of adherence to the WBS linked with the capacity of the WFP software to enhance data integrity will facilitate this process: See Dashboard flow chart.

77. *Develop FIWPs:* Construction Contractor: The WFP Model Administrator will apply the CWP (Part A) to the model and utilize the WFP software to facilitate the dissection of the CWP into FIWPs by the WorkFace Planners. The grouping of drawings into FIWPs (in Excel) is a deliverable from this session to the Material Management Team. The list of FIWPs along with the sequence and durations are a deliverable from this session to the Construction Contractor's Scheduler. : See Basic Principles – Schedule for Sale.

78. **Apply Virtual FIWPs to MMD:** Material Management: The MMD administrator will receive a list of drawings and components by FIWP (from activity # 77). This information along with the preferred sequence will be added to the MMD so that materials can be electronically grouped by FIWP and that a priority sequence can be established for dynamic allocation.

79. **Develop** <u>level-5</u> **Schedule:** Construction Contractor: The Scheduler will enter the individual FIWPs as <u>level-5</u> activities based upon the information generated during activity #77. This activity can take place at the same time if the scheduler is present during activity #77.

80. **Bag and Tag Materials by FIWP:** Material Management: The Warehouse Team for the Material Management Team receives a "material notification" as the FIWP is entered into the four-week look ahead schedule. The material is preassembled based upon the estimated ROS (Required On Site) date and is then delivered once the "material request" is received.

81. **Remove Constraints:** The WorkFace Planners take the virtual FIWPs created during activity # 77 and enter them into <u>Pack Track</u>. The WorkFace Planners apply constraints against each FIWP and then work on removing the constraints. When the material for a specific FIWP is confirmed received then the FIWP is developed into a hard copy and moved into the four-week look ahead schedule. The removal of constraints continues until the FIWP is free of constraints and ready for release. The cycle length for this activity is 8 weeks, based upon the minimum fabrication cycle for steel and pipe.

82. **Execute FIWPs:** Construction Contractor: Working from the four week look ahead schedule (a portion of the <u>level-5</u> project schedule) the General Foremen collects the FIWPs in the week prior to execution and distributes them to the Foremen. The Foremen confirm the work scope and scaffold erection, then execute the work from the start of the next rotation (Monday morning).

83. **Record Progress:** Construction Contractor: The Foremen report progress against the FIWP daily to the General Foreman. The GF then reports progress to the WFPs

who enter it into the WFP software. The Project Controls Team utilize the WFP software to roll up progress once a week and report against level-3 schedule activities to the PMT. See WFP Project Controls

84. **Record Progress:** Project Management: The Project Controls team for the PMT receive progress from the Construction Contractors. The difference for the application of WFP is that the Project Controls Team (PMT) have view only access rights to the WFP software and model, so as the progress is recorded the progress changes daily and presents a live 3D model of exactly what is being progressed. The contractor will continue to meet weekly with the PMT to present a summary of the progress and performance report, against level-3 schedule activities.

Note: The alignment between schedule activities and the 3D model components also enables the 4th dimension (time). This allows model administrators to play a sequencing video of the FIWPs as they are scheduled.

Close: Turnover

85. **Monitor and Control:** Project Management: The Project Management Team manages the turnover deliverables against the plan.

86. **Maintain Model with Red Line Changes:** Engineering: As the Engineering Team move to site to support Construction they will also be required to maintain the model with revisions and redline asbuilt drawings. This maintains the model's integrity and prepares the model for turnover to RFO and Operations.

87. **Manage Documentation for System Turnover:** Engineering: The Document Control Database will be structured to store and display all of the documents required for turnover. As templates are created for testing, punchlists and turnover the Construction Contractor will submit the soft copy to Document Control and then have the ability to manage and edit them online. The hard copy exhibit documents will be

scanned (PDFs) and submitted to the Turnover section of Document Control as they are created.

88. **Develop Test Packages:** Construction Contractor: The WFP team transition from building construction FIWPs to developing Test package FIWPs for hydro testing and loop checking. The WFP software will be utilized to display the model by system and then to apply test boundaries. The actual progress can then be applied to test packages and be used to complete construction in preparation for testing.

89. **Develop Turnover Packages:** Construction Contractor: The WorkFace Planners transition from developing test packages to developing turnover packages under the direct supervision of the Construction Contractor's Turnover Coordinator. The WorkFace Planners utilize the WFP software to isolate systems and build 3D snapshots that show the systems: vessel to vessel or start to finish. The turnover FIWP also contains a copy of the QC documentation and sign off docs for transfer of Care Custody and Control. The WFP software can then be utilized to display the systems that have been turned over.

90. **Develop Construction Punchlists:** Construction Contractor: Immediately prior to the transition from bulk construction to systems construction ,the WFP Team will transition from building test packages to building punchlists based upon turnover systems. The WFP software can be utilized to isolate systems and then display the work that still needs to be completed. This can then be rolled up to form a % complete for each system. The master punchlist should be maintained live in Document Control.

91. **Submit QC docs (ongoing):** Construction Contractor: The QC group will collect documents from the completed FIWPs and scan and submit them daily to the Document Control Database. Original documents are bar-coded and stored. The WFP software can monitor the process of documentation management for system turnover.

92. **Apply Third Party Audits:** Owner: The owner will arrange for and coordinate third party audits as per their plan

established in Activity # 27. The results from the audits will be shared with all of the stakeholders and the Project Management Team. The timing of this audit should be soon after the transition from Bulk construction to System construction. This will ensure that the processes are in place for the WFP facilitation of system turnover.

93. **Turnover the Model Systems at MC:** Engineering: The Engineering Team are expected to maintain the 3D model with Redline changes so that the model can be turned over to operations. The model will be turned over one system at a time as part of the system turnover deliverables upon Mechanical Completion (MC). This will facilitate the WFP support for the Ready for Operation and Commissioning & Start Up Teams.

94. **Coordinate Turnover:** Construction Contractor: Typical to most projects. The only variance is that the Turnover Coordinator will be able to utilize the WorkFace Planners, the 3D model and WFP software to develop, monitor and report Turnover FIWPs.

95. **Coordinate Turnover:** Project Management: This is an oversight role and an existing activity that is typical to most projects. The difference will be the Turnover Coordinator will be able to monitor system readiness through the WFP software and progress reports from Document Control.

96. **Mechanical Completion – System Turnover:** Mechanical Completion – System Turnover: Milestone for all parties – standard practice.

97. **Support RFO and C&SU with WFP and Model Management:** Construction Contractor: The Construction Contractor will make available the WorkFace Planners, the 3D model administrator and the WFP software to the Ready for Operation and Commissioning & Start up Teams. This will allow the Operations groups to utilize the existing information for education or to build plans and develop simulations.

98. **Accept Systems (Ready For Operations)** Owner: RFO team accept Care, Custody and Control of a system.

99. **Prepare Systems for Start up (Commissioning and Start Up):** Owner: RFO team hand over a system

to Commissioning and Start up in preparation for operation.

100. **Record WFP Lesson Learned:** Project Management: The WorkFace Planning Manager will lead the WFP Stakeholder Team through a half-day lesson learned meeting, where the team will revisit each stage of the project and record the lessons learned. The WorkFace Planning Manager will then formalize the meeting minutes and lessons learned and distribute the final document to all of the Stakeholders and their organizations.

101. **Record Lesson Learned:** Project Management: A review of the entire project conducted by the Project Management Team.

The web site www.scheduleforsale.com contains a series of documents that support the application of "WorkFace Planning by Design". Members can download these documents or print them directly from the web page. The links built into the flow chart nodes and will take you directly to the relevant documents or you can find them under the Support Documents Tab.

Chapter 11

Total Information Management

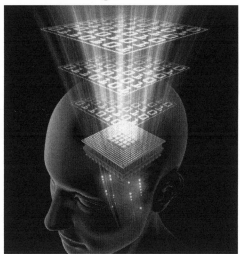

Data → Information → Knowledge → Understanding

Introduction:

Men occasionally stumble over the truth, but most of them pick themselves up and hurry off as if nothing happened.

Winston Churchill.

After many years of study and billions of dollars wasted in a frustrating effort to "get those guys working", the industry is finally waking up to the fact that we are the problem. This truth is not very convenient. It would be much simpler to blame labour for poor productivity and then build an incentive plan that would whip them into shape. After all that is what our Fathers and Grandfathers did and it seemed to work for them.

The game has changed and the rules must change to suit the new game.

Think back 50 years and try to imagine a Foreman for a crew of Carpenters, (Your Grandfather?) now think forward 20 years and think of the same position (Your Son/Daughter). Which skills changed, which skills disappeared and which new skills appeared?

Your answers will lead you to your new job description.

The truth is that if you are reading this book then you are now, or soon will be, in a position to change the way that we do business. Your mandate is to initiate the changes that will prepare the industry for the next generation of project management. This will allow our Sons and Daughters to be the Supervisor of the future that you just imagined.

So what were the key differences that you saw between the past and the future?

I would like to guess that several of the changes were centered upon information management, amongst other things. My own vision of the next supervisor's tool box includes tools that process data and answer questions through web links, accessed through hand held computers. This is not a stretch from where we are today. The sad truth is that these tools are available now, we just don't have the processes that are needed to produce the clean data for input.

Therefore, our goal is not to build the tools but to develop the processes.

A simple example of this is steel piece mark numbers.

Our engineering teams will typically design steel structures without the connection details or the unique identifiers for each piece. The detailers who work for the fabricators add the connection details and the piece mark numbers. Both of these teams use software to achieve this but the information is transferred between them on paper through the mail. The result is that the constructor receives a picture from Engineering (the 3D model) and a set of drawings issued by the detailer. On the good projects there is a 95% match and it will only take the foreman a day or two to correlate the picture with the drawing and the truckload of steel pieces.

With a slight change to the existing processes, we could give the Foreman printed 3D images that have all of the piece mark numbers on the steel and then match that with a truck loaded with all of the right pieces, in the right order.

The Foreman would then be able to spend more time with the crew and the crew would be able to access the right steel in the right sequence. This would lead to productivity improvements for the steel erectors and also for all of the other trades that rely on the erection of steel for their work fronts.

And that is why **Total Information Management** is the true path to productivity improvement.

So how does this connect with WorkFace Planning?

Every time we apply WorkFace Planning, the search for information leads us upstream to the information generators: Engineering, Procurement, Project Controls, Document Control or Material Management. The root of the problem is always familiar: The information was generated but not passed on, passed on in a format that could not be applied or deleted because nobody asked for it.

WorkFace Planning gives our organizations a target for information. In one memorable example, a Senior Project Manager held up a FIWP at a weekly project team meeting and said, "Your job description is to make this plan workable." This was His way of saying that the information that our teams generate has to make it to the finish line, it has to support construction.

This example of steel engineering and fabrication is one tiny piece of the puzzle. Every process that we look at in construction has opportunities for process development. The spark that will start this revolution is WorkFace Planning. The road map that will show us where to go is the 3D model and WorkFace Planning Software.

WorkFace Planning Software:

We have now had glimpses of what Total Information Management might look like, thanks to several well-organized projects where WorkFace Planning Software was applied.

The early adaptations of WorkFace Planning Software through Fluor (InVision), KBR (P6) and Jacobs (ConstructSim) have all shown us that the right information in the right hands can bring extraordinary results.

These three software applications and the ones in development by Intergraph and Aveva are all based upon the logic that the 3D model can be used to facilitate the process of WorkFace Planning and then visualize the information.

A simple example of this is to take a list of received material from the laydown yard and then select the same objects in the model. The software then groups the objects and makes them all bright green in the model. The WorkFace Planners can then look at the model and understand exactly what material is available.

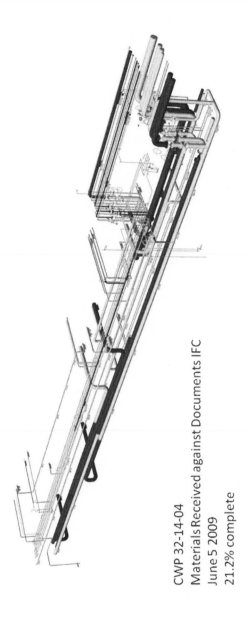

CWP 32-14-04
Materials Received against Documents IFC
June 5 2009
21.2% complete

The same logic can be applied to documents, progress and the schedule (The 4th Dimension). The model then becomes a wonderful communication tool that takes data and transforms it into information.

I should mention at this point that I have never been on a project that has reached the point where all of this information was accurately displayed in the model at one time. The source of the problem was and still is: clean reliable data. As harped on in earlier chapters the size and shape of data must be designed to be fit for purpose, at the start of the project.

The leap from Information to Knowledge occurs when the WorkFace Planner interacts with the information displayed in the model. This is where the background experience of the WorkFace Planners becomes very important. A colour-coded model is a wonderful source of information, but knowing what to do with it is the leap to Knowledge.

This is also the next leap for WorkFace Planning software. Most of the software applications allow the administrators to install unit rates as a reference database. This is usually a standard platform of known installation rates for commodities (4 hours per foot of 6" pipe) with site-specific multipliers, that can be proprietary or off the shelf (Page & Nation's Estimator's piping manual). The software then multiplies a chosen object's length or weight by the supplied unit rates. This tells the WorkFace Planner what the standard expectation is for performance (Planned Value). The WorkFace Planners use this information as a backup to their own interpretation of the capacity of a work crew based upon their understanding of the current reality. It is important that the WorkFace Planners design FIWPs primarily based upon their own expectations with secondary consideration for installation rates. (This allows us to over-perform).

The WFP software then allows the WorkFace Planners to interact with the model to group objects. The users will click and drag objects into a plan. The objects don't move but the plan is populated with the details of the object and the object in the model becomes a member of a group (The FIWP). The colour of the objects change as they are added to groups and the WorkFace Planners can then work their way through a structure until all of the components are coloured (planned). The software stores the grouping of the objects. These groups can then be checked in the model against the constraints of documents and materials.

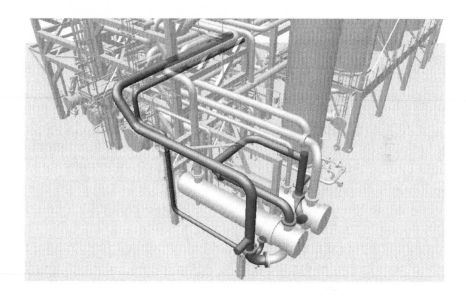

The groupings are then branded with their FIWP naming convention and sequenced with proposed start dates. The WorkFace Planning software will then allow a playback based upon dates, this creates the 4th dimension (time). Imagine this steel CWP appearing on the screen FIWP by FIWP. When you play this back to the constructors they add further scrutiny to the sequencing and start to identify construction constraints.

Each FIWP then appears as a level-5 schedule activity. The end result is a real level-5 schedule based upon construction logic.

And this is the leap from Knowledge to Understanding. We have now taken a conceptual idea (the schedule) and created concrete comprehension in the minds of the people who are responsible to do the work. At the same time we have fed concrete understanding back into the concept. (Level-5 schedule activities). The result may well be that the original concept is forced to change based upon our newly developed concrete understanding.

If the original window of execution for a level-3 activity (CWP) was two months and the dissection of the schedule into FIWPs shows that it cannot be performed without additional resources or a second shift, then the concept may be forced to change. That is the leap from knowledge to understanding, drawn from experience and reality.

You should be having an uh-ha moment soon.

The WFP Software captures and stores the allocation of components to FIWPs. An extract of this data can then be uploaded into the material management database. This allows materials to be received electronically against their FIWP, with dynamic allocation of bulks based upon FIWP start dates.

So at this point, we can plot the materials and documents received against a whole CWP of FIWPs, eight weeks prior to the closest start date. The process of constraint removal then proceeds and we can track the progress in the WFP software.

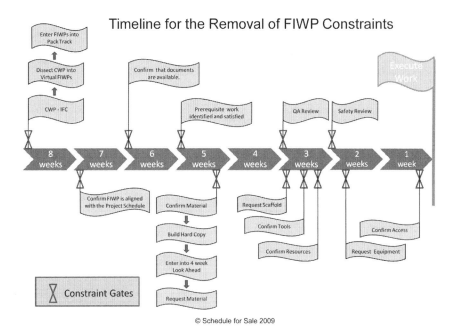

Timeline for the Removal of FIWP Constraints

© Schedule for Sale 2009

Each FIWP marches through this list of constraint removals (from Chapter 2) until it is released to the field and executed. At this point, there are many more processes that are developed for each project to ensure that the work can be executed as planned. These include how and when the material is delivered to site, how is it received, how do you address RFI's, how many FIWPs get released to the GF at a time and how the timesheets are cost coded to align earned hours with burned hours. All of these processes and a series of others need to be developed project by project based upon the way you like to execute work. The target is the successful execution of FIWPs within their scheduled window of time.

The Key Performance Indicator (KPI) from the execution of FIWPs is progress. How did we perform against our expectations? This closes the loop on all of our hard work from the start of the project.

Today's WFP software is wonderfully equipped to manage progress. The impact of visualized progress is so powerful that it will dramatically change the way we think of project controls in the very near future.

Some of the WFP software will develop a check sheet that lists each component and each stage of development, (receive, install, connect and test) this should be included in each FIWP. As each FIWP is executed, the Foremen will utilize this check sheet to mark and report daily progress to their General Foremen and/or the WorkFace Planner. The WorkFace Planner highlights the FIWP stored in the WFP software and updates the status of each component. The software then calculates the earned value and % complete of each FIWP.

It is the General Foreman's job to monitor day-to-day progress and they may utilize the WFP software to do this. However, care must be taken to not draw conclusions from the lack of early progress earned. Many tasks will take more than one day to complete so will not earn value until the second or third day.

The Project Controls Team can draw progress from the WFP software at the end of the rotation in the format that they prefer, usually earned value in hours rolled up to CWPs. The WFP software will also produce a 3D snapshot of the reported progress, which is an excellent way to show <u>what</u> is getting done alongside <u>how much</u> got done.

At this time WorkFace Planning software does not calculate productivity factors (earned hours/burned hours). This is still managed by timesheet collation software and a project controls database, with the correlation usually rolled up to a CWP level. If you can get the timesheet system to spit out how many hours where charged to each FIWP you could use excel to quickly calculate a PF for each FIWP. This is useful if you want to have a closer look at a CWP that is falling behind schedule.

In the years to come we can expect to see WFP software develop to also contain scaffold and equipment management processes, RFI tracking logs, QC management features and engineering management processes. Each of the applications also has other features, not directly related to the application of WorkFace Planning. You can visit the websites listed on the last page to get more information.

As an end User, I am very happy to have such fine products available to use and am excited at the constant development of this technology. However, this does come with a note of caution: The software does not perform miracles, (I know, I have been on projects where nothing short of a miracle was enough). The software is very good at showing you your current reality. This is not always welcome news and leads to finger pointing that invariably leads to the questions about the integrity of the software. At the end of these witch-hunts, you will find that the software visualizes your data, and that is the root of the problem. The organizations that will leap to the forefront of our industry and become the premier suppliers of services will be the ones that use this harsh reality of truth to focus their efforts, not to lay blame.

The previous couple of passages may not sound like much but this is what it is all about. This is where the process and the tools come together. The process is as important as the tool.

WorkFace Planning needs WorkFace Planning software and WorkFace Planning software needs WorkFace Planning.

If you get to the stage where you have WorkFace Planners, FIWPs, a process for removing constraints and WorkFace Planning software then your construction productivity will improve dramatically. Your team's inertia will gradually pick up and your job will change from trying to get people to do stuff to a position where you work for them. Your job will now be to run interference for your team by removing obstacles before they get in the way. Get ready to say "Yes" loud and often, when your team asks for stuff.

And that is the basic difference between Managers and Leaders, Managers work on trying to maintain the status quo, while Leaders find a way to say "Yes" to change. Our success needs both Managers and Leaders. Leaders that will take us to new places (WorkFace Planning) and Managers that can stabilize and perfect the systems when we get there.

If you choose to go down the road of WorkFace Planning without an application of WorkFace Planning software (the tool) you will be on a well travelled path. Most organization have to get it wrong before they understand the true value of doing things right.

However, as mentioned once before in this book, you will not live long enough to make all of the mistakes that you have to, so teach yourself how to learn from the mistakes of others.........

: Get the software.

Chapter 12

Beyond WorkFace Planning and WorkFace Planning Software:

The following and final chapter of the book is a look into the crystal ball of the next 10 years. Some of it is easy to see. When you get your head around ISO 15926 you will be embarrassed by the fact that we don't already do this. Other stuff is a little more out there on the edge of what is possible.

Ultimately, our projects <u>will</u> get to the point where they are information centric. The primary product of our processes is information, right up to the point where we fabricate something or construct it. So gravity is moving us all towards this common logic of WorkFace Planning and the Total Management of Information.

The following diagram is a broad look at the stakeholders and the processes. Each department is both supplier and a consumer of information. Looking at this business model from a distance it makes good sense to predetermine the life cycle of information and then design processes that will produce information that is fit for purpose. The tools are already out there to do this we just need better understanding of the questions.

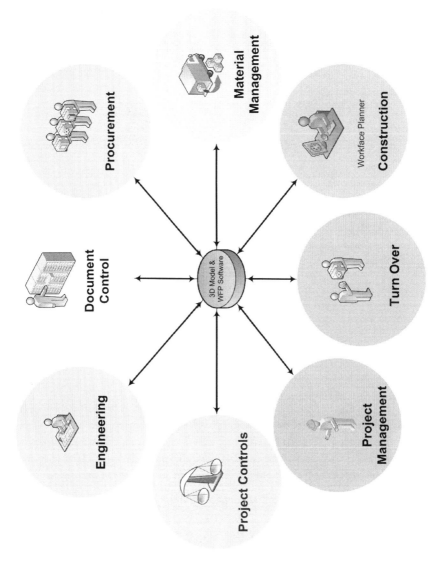

The ultimate outcome is to develop a 3D model that answers all of the project's questions around scope, schedule, costs, materials, engineering, documents, progress, and turnover.

Imagine a project where everybody accessed an online 3D model for information and reports on all of the project's Key Performance Indicators (KPIs).

The model administrators would monitor the weekly inputs from each department and apply the data through the WorkFace Planning software. Everybody on the project would be able to answer most of their questions from their own desktop.

It may look something like this:

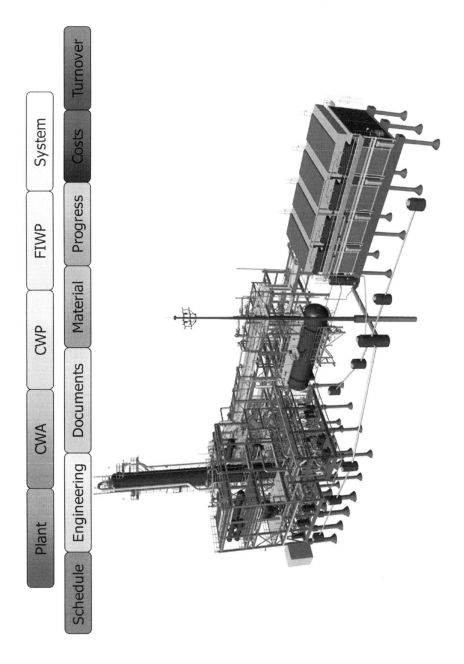

Plant | CWA | CWP | FIWP | System

Schedule | Engineering | Documents | Material | Progress | Costs | Turnover

As you mouse over the menu items, a detailed menu appears and you make another choice. First for the size and shape of the image (CWA, CWP, FIWP or System) and then for the information that you want to see. The model colours the components and shows you an explanation of the colour coding.

The link to this nirvana is a clean, consistent supply of data.

"Technology is waiting impatiently for process to catch up." *(Schedule for Sale)*.

One of the wonderful things happening in the world of data management is the development of the ISO standard 15926. The standard prescribes the data content and format for all components used in oil and gas production facilities so that the information can be utilized by different software applications. This allows us to expect full life cycle data management, exchange and integration.

The world's leading software developers have accepted the standard, initiated in Europe and then developed across the whole world through open source, as the established benchmark for all future data generation and management.

For Owners and Project Management teams it is a very good fit with the logic of Total Information Management. You can simply add this standard to your contacts for all project stakeholders and then manage to the standard. You would probably have to start by developing your own inhouse understanding of 15926 and then be willing to work with your Engineers, Vendors and Fabricators to help them mature to this new level of data integrity. However, the end result far outweighs this initial investment.

Imagine a 3D model that is fully loaded with standardized information that can be used to design, procure, fabricate, construct, test, turnover, start up, commission, operate, maintain, and decommission a plant. That is full life cycle data management and we already have the technology to do it, we just need the data.

An interesting observation made by an oil company executive was that the application of ISO 15926 means that the owners can now own and utilize engineering data that they have purchased. This allows them to apply it to a whole host of future applications and any software that they choose. They own the data. In the absence of 15926 Owners have to purchase the specific software that was used to generate the data and probably employ the company that produced the information.

Please have a look yourself at : http://en.wikipedia.org/wiki/ISO 15926 . The level of detail is one or two steps removed from my level of comprehension but I do understand the logic. The simple explanation (given to me) was that it is like having a library that lists everything that you could use to build a petrochemical plant. Then each item has a series of blank boxes that must be populated with information, in the xml format.

This standardised information can then become the basis for a project management system that targets the efficient exchange of data. The iRING (ISO15926 Real-time Interoperability Network Grid) project developed by FIATECH is a good example of this (http://fiatech.org/press-releases/280-peer-to-peer-iring.html). A peer to peer Infrastructure based upon interoperability using data that is ISO 15926 compliant. The example that we witnessed at the FIATECH conference in 2009 showed data being exchanged in seconds between three engineering companies across four continents and several software applications. (This infrastructure is free to any organization that would like to use it). If the operation of a system like this is made possible by ISO15926 imagine what it could do for your project. (work=time=cost)

The application of ISO15926 has also had a positive impact on the application of BIM (Building Information Modeling) in the building construction industry. (http://en.wikipedia.org/wiki/Building Information Modeling). The application of BIM is being requested by more and more building construction owners and could be viewed as a model for the Total

Information Management model that I described at the start of this chapter.

In 2004 the US National Institute of Standards and Technology released a report that estimated more than $15 billion was lost each year due to the inadequate management of information in the building construction industry. The report listed "inconsistent technology adoption among stakeholders" as a major contributor to this problem.

Conclusion:

Sounds big and impossible? Let's take a moment and look into the future from the past.

When we compare the Gross Domestic Product of Canada to the world, we see that it takes 99 people in Uganda to produce the equivalent of one person's production in Canada. (Both countries have about 33 million people) It wasn't always this way, Canadians use to produce a lot less 100 years ago. And it is probably not that Canadians work 99 times harder than the good folks of Uganda. There is a good chance that it is because Canadians have equipment and processes (technologies) that allow them to produce well. So if the ranking on the scale of Gross Domestic Production (productivity) is less about effort and more about technology then it makes sense that the more technology we use the higher our productivity becomes.

New Technologies = Increased Productivity

This principle is constant and applies at all levels of application.

So if we truly want to continuously improve our performance in the field of Project and Construction Management then we will need to keep introducing new technologies:

WorkFace Planning and WorkFace Planning software.

I do know now that "impossible" is just a temporary state of mind, you'll get over it, like I did.

Geoff Ryan

"Never tell a person that anything cannot be done.

God may have been waiting centuries for someone ignorant enough of the impossible to do that very thing."

GM Trevelyan

Some Resources:

Schedule for Sale webiste: www.scheduleforsale.com

Insight-wfp.com –Consulting, Coaching and Application provider for WorkFace Planning .
www.insight-wfp.com

Rally Engineering Inc: WorkFace Planning based Project Management and Engineering Company.
www.rallyeng.com

ASI – Consulting and Training provider for WorkFace Planning
www.ascensionsysytemsinc.com

WorkFace Planning Institute: Not for profit community of WorkFace Planners. www.workfaceplanner.com

Construction Owners Association of Alberta (COAA)
www.coaa.ab.ca

COAA Best Practice Model for WorkFace Planning.
www.workfaceplan.com

Aveva net: WorkFace Planning Software provider.
www.aveva.com

Bentley: WorkFace Planning Software provider.
www.bentley.com/en-US/Products/ConstructSim

Intergraph: WorkFace Planning Software Provider.
www.intergraph.com/ppm/spc.aspx

Edwards Brothers Malloy
Oxnard, CA USA
August 8, 2014